IMAGES
of America

HISTORIC
JOURNEYS
INTO SPACE

IMAGES
of America

HISTORIC
JOURNEYS
INTO SPACE

Lynn M. Homan and Thomas Reilly

ARCADIA
PUBLISHING

Published by Arcadia Publishing
Charleston, South Carolina

Printed in the United States of America

Library of Congress Catalog Card Number:

For all general information contact Arcadia Publishing at: Applied for
Telephone 843-853-2070
Fax 843-853-0044
E-mail sales@arcadiapublishing.com
For customer service and orders:
Toll-Free 1-888-313-2665

Visit us on the Internet at www.arcadiapublishing.com

CONTENTS

ACKNOWLEDGMENTS

Hundreds of thousands of photographs have been taken during America's exploration of space. Both amateur and professional photographers have captured historic moments on Earth, up to and including liftoff. From that point, astronauts have used handheld cameras to document particular aspects of a mission, while automated cameras on board have recorded all of the myriad details of a space flight.

The National Aeronautics and Space Administration (NASA) has made its wonderful collection of photographs, films, and other images available to the world. Without the cooperation and assistance of the men and women who work in the photo archives at the various space centers, however, access to this tremendous resource would be severely curtailed. We are grateful to everyone who provided assistance but most especially, the following individuals: Jody Russell and Debbie Dodds at Johnson Space Center, Margaret Persinger at Kennedy Space Center, Constance Moore at NASA Headquarters, and Judy Pettus at Marshall Space Flight Center.

All of the photographs used in this book have been obtained from the National Aeronautics and Space Administration and are hereby credited accordingly.

INTRODUCTION

The countdown had begun—10, 9, 8, 7. On April 12, 1981, millions of people waited in anticipation—6, 5, 4, 3. With tightened muscles in the backs of their necks, American television viewers stared blankly at their television sets. Another disappointment would not be easily accepted; there had already been too many. Only two days earlier, the scheduled launch of America's first space shuttle mission, STS-1 *Columbia*, was scrubbed because of a computer problem. The sight on the television screen was overwhelming. *Columbia*, named after an 18th-century Massachusetts-based sailing ship, was the size of a modern jetliner. Its empty weight was almost 200,000 pounds. Two men, John W. Young and Robert L. Crippen, sat inside *Columbia*'s cockpit waiting for liftoff from pad 39A at Kennedy Space Center. There was a great deal riding on this mission. This reusable spacecraft was the culmination of a $9 billion investment. It was the first American space mission in six years.

The countdown continued—2, 1, liftoff. At only a few seconds past 7:00 a.m. eastern standard time, America's first shuttle mission was launched. *Columbia* would soon be 166 miles above the Earth. During the two-and-one-quarter-day mission, *Columbia* traveled a distance of 1,074,567 miles and orbited the Earth 37 times. A tremendous success, *Columbia* demonstrated a safe launch into orbit and the successful return of a reusable spacecraft—the space shuttle.

The aerodynamically beautiful space shuttles that Americans have come to expect as a new millennium dawns were a long time in the works. For years the race for conquest of space had gone on between America and the Soviet Union. It was ironic that one of Adolph Hitler's most feared weapons of World War II, the V2 rocket, would have been the beginning vehicle of America's space program. At the end of the war, 100 German V2 rockets had been transported to the United States and were eventually configured for the launch of the earliest unmanned missions. Launch of the Soviet Union's artificial satellite, *Sputnik I*, on October 4, 1957, pushed the American space program out of the doldrums. With wounded patriotic pride, Americans feared for their national security as the cold war raged. An American satellite was launched into space in January 1958. The space race was on in earnest.

Created by President Dwight D. Eisenhower, the National Aeronautics and Space Administration (NASA) was announced on October 1, 1958. One week later, on October 7, Project Mercury was approved. Its mission was to send the first American into orbit. Six months later, a press conference was held in Washington, D.C. The date was April 9, 1959, and the first American astronauts, the Mercury seven, were presented to the American public. The selection criteria required that they be less than 40 years old, less than 5 feet, 11 inches in height, in excellent physical condition, and have a test pilot background with at least 1,500 flight hours. Nearly 1,000 American men applied for these few precious slots. The select seven included Walter M. Schirra Jr., Donald K. Slayton, John H. Glenn Jr., Malcolm Scott Carpenter, Alan S. Shepard Jr., Virgil I. Grissom, and L. Gordon Cooper Jr. The goal of the Mercury project was to put an American into space before 1960 came to an end.

Only 20 years separated the launches of the first American shuttle and *Mercury 3*, the first American manned space flight. Launched atop a Mercury-Redstone rocket on May 5, 1961, Alan Shepard became the first American astronaut in space. As dramatic as the 15-minute, 22-second *Mercury 3* launch had been, America was still playing catch-up with the Soviet Union. Soviet cosmonaut Yury Gagarin, on April 4, 1961, had earned the distinction of being the first man in space.

President John F. Kennedy presented Alan Shepard the Distinguished Service Medal on May 8, 1961. On May 25, before Congress, Kennedy requested funding for a mission, "before this decade is out of landing a man on the Moon and returning him safely to Earth." The gauntlet had been thrown; Kennedy wanted to beat the Soviets to the Moon. By the mid-1960s, six Mercury launches would put six Americans into space at a cost of $392 million. Donald Slayton was the only one of the original seven Mercury astronauts who did not have the opportunity to fly a Mercury mission.

In December 1961, NASA announced plans for a two-man spacecraft that was to be named *Gemini* after the mythological heavenly twins. *Gemini* was intended to be the foundation by which to gain the necessary experience to allow completion of NASA's goal—a manned mission to the Moon. Instead of an Atlas rocket, a two-stage booster Titan intercontinental ballistic missile was to be used for launches.

Gemini 3, the first manned Gemini mission, took place on March 23, 1965, with astronauts Gus Grissom and John Young riding on board a two-man capsule nicknamed *Molly Brown*. Only three months later, on June 3, James McDivitt and Edward White were launched into space. In a four-day *Gemini 4* mission, Edward White became the first American to walk in space. Speeding above the Earth at 17,500 miles per hour, while attached to a 27-foot-long tether, White walked in space for 21 minutes. In a total of ten missions, the Gemini program accomplished what it had set out to do. Long periods of weightlessness had been experienced, precision docking of a pair of spacecraft had been accomplished, and man had walked in space. Gemini had propelled the United States far ahead of the Soviet space program; it proved that the United States could send a manned mission to the Moon.

The three-man Apollo project was designed to take an American to the Moon. Safety was obviously of paramount importance. The decade of the sixties had not been kind to America's astronauts. No American astronaut had yet died while on a mission. However, four astronauts had lost their lives. Theodore Freeman, an astronaut for a year, died in October 1964, when his T-38 airplane hit a flock of geese. Gemini crewmen Elliott See and Charles Bassett also died while flying a T-38 in 1966. Edward Givens perished in a car crash on June 6, 1967, in Houston, Texas. Almost as if jinxed, a T-38 jet took the life of yet another astronaut when Clifton Curtis Williams died on October 5, 1967.

The first Apollo mission, *Apollo 1*, originally scheduled for December 1966, was intended as a shakedown flight. The scheduled two-week mission seemed doomed from the beginning. Constant problems beset *Apollo 1*. The launch date was rescheduled for February 1967. Ground testing was slated for January 27. Gus Grissom, Ed White, and Roger Chaffee had already been in the capsule for nearly six hours as testing took place. At 6:30 p.m. in the darkness of a Florida winter night, a fire broke out in the capsule. Roger Chaffee called out, "We've got a bad fire . . .We're burning up here!" Sixteen seconds after Chaffee's cry, the capsule burst into flames. With the destruction of *Apollo 1* and the deaths of Grissom, White, and Chaffee, America's lunar program was set back at least two years. President John Kennedy's wish for America to be first on the Moon looked doubtful.

Not until October 11, 1968, did America finally witness a manned Apollo mission. Former

Mercury astronaut Wally Schirra, Donn Eisele, and Walter Cunningham successfully completed a 4.5-million-mile flight. Once again, the quest to fly to the Moon was on. Within the next seven months, another three Apollo missions were flown. Each mission paved new ground and tested new equipment.

"Houston, Tranquillity Base here. The *Eagle* has landed." Few will ever forget those words spoken by Neil Armstrong on July 20, 1969. America had done it. The Moon had been conquered. Less than seven hours after the lunar module *Eagle* carrying Neil Armstrong and Edwin "Buzz" Aldrin landed on the Moon, it was time to set foot on another world. As Armstrong stepped onto the surface of the Moon, he said, "That's one small step for man, one giant leap for mankind." Soon, Aldrin joined Armstrong. Two Americans had now walked on the Moon.

Six more Apollo missions traveled to the Moon before the program ended with *Apollo 17* in December 1972. Gene Cernan and Harrison Schmitt became the last two men to walk on the lunar surface. Apollo had lasted more than four years, three Apollo astronauts had died, and a dozen men had walked on the Moon. America had clearly won the space race.

Following the successful Apollo missions, four Skylab missions were launched between May and November 1973. NASA's goal was a low-cost, reusable, orbital space station. An unmanned 85-ton *Skylab* workshop was launched on May 14. The workshop boasted 10,000 cubic feet of space and a docking adapter; it was provisioned with air, water, clothing, and a ton of food. The luxurious quarters offered sleeping bags, showers, and recreation facilities. Before *Skylab* crashed to Earth on July 11, 1979, it had been used by the crews of *Skylab 2, 3,* and *4.*

Reusable space shuttles were the next logical progression in America's space program. America's first space shuttle orbiter, *Enterprise*, was rolled out of Rockwell's Air Force Plant 42 in California on September 17, 1976. As would be all of the future shuttles, *Enterprise* was originally supposed to be named after a sea vessel. A name change resulted after fans of the *Star Trek* television show waged a successful write-in campaign asking the White House to name the shuttle *Enterprise* instead of *Constitution. Enterprise* had been a tremendous undertaking. The contract, which was first awarded in July 1972, required the manufacture of millions of parts by hundreds of thousands of aerospace workers. *Enterprise* was used for ground and air testing and never flew in space. In its wake followed five additional space shuttle orbiters: *Columbia, Challenger, Discovery, Atlantis,* and *Endeavour.*

Following the completion of *Columbia*'s first flight in 1981, John W. Young, STS-1's mission commander, said, "A really fantastic mission from start to finish. The human race is not too far from the stars." It seemed as if nothing could stop America's 20th-century quest for manifest destiny. Each year, as the number of shuttle missions increased, additional barriers were broken. Sally Ride, the first American woman in space, was a member of the *Challenger* STS-7 mission. Guion Bluford became the first African-American to fly on an American spacecraft as part of *Challenger* STS-8. Space travel had become routine, safe, almost boring. Even politicians had flown as passengers. Civilian mission specialists had become members of shuttle crews. Sharon Christa McAuliffe, a teacher, was to serve as a payload specialist on shuttle *Challenger* 51-L. During the scheduled seven-day mission, McAuliffe planned live telecasts during which she would teach from space.

The boring routine of shuttle launches ended on January 28, 1986. Only 73 seconds after launch, *Challenger* had reached an altitude of 46,000 feet and a speed of nearly Mach 2. In sight of thousands of spectators at Cape Canaveral and millions more watching a live television broadcast, the unthinkable happened. *Challenger* exploded. The crew of seven perished. With 15 flights scheduled, 1986 was to have been the year of the shuttle. Missions did not resume for another two years.

Since 1988, America's space shuttle orbiters have flown nearly 100 flights and logged millions of miles and several hundred days in space. Other important milestones have also occurred. On STS-47, Mae C. Jemison made history as the first African-American woman to serve as a NASA mission specialist. In 1998, John H. Glenn Jr., the first American to orbit the Earth, returned to space at the age of 77 to become America's oldest astronaut. Just a few months later on STS-93, Eileen M. Collins became the first female commander of a shuttle mission.

Poised at the beginning of a new century, America's space program prepares to continue its historic journey. Scientific experiments produce exciting new data and discoveries. The deployment of the Hubble telescope and the Chandra X-ray observatory allow scientists to gather important data, making the exploration of distant planets increasingly possible. Unmanned spacecraft have landed on the surface of Mars. Cameras send back never-before-seen images from deep in the universe. Just as recent accomplishments were once unimaginable, one can only guess as to the new frontiers still to be conquered by America's space program.

One

PROJECT MERCURY
INTO SPACE

Launch personnel in the blockhouse at Cape Canaveral had thousands of details to attend to before the launch of the *Explorer 1* satellite on January 31, 1958. The first artificial satellite to be launched by the United States, *Explorer 1* weighed only 30.8 pounds and circled the Earth in 1 hour and 55 minutes. The Soviet Union's *Sputnik I,* launched on October 4, 1957, was the world's first artificial satellite to enter space. Barely 2 feet in diameter and weighing 184 pounds, *Sputnik I* attained a height of 584 miles above the Earth. Only a month later, *Sputnik II* was launched; its payload was a dog that lived for four days. With the launch of the pair of *Sputniks,* the Soviet Union had clearly taken the early lead in the space race. Now America's Project Mercury was on its way.

A Juno I rocket sits on a launch pad at Florida's Cape Canaveral in anticipation of its launch on January 31, 1958. The Juno I rocket, a reconfigured Jupiter-C rocket, served as the launch vehicle for *Explorer 1*.

The original seven Mercury astronauts pose for posterity in their space suits. They are, from left to right, as follows: (front row) Walter Schirra Jr., Donald K. Slayton, John H. Glenn Jr., and Malcolm Scott Carpenter; (back row) Alan B. Shepard Jr., Virgil I. Grissom, and L. Gordon Cooper Jr. These first American astronauts were presented to the American public in April 1959. NASA's goal for the Mercury project was clear; one of these seven Americans would be launched into space before 1960 came to a close.

NASA's goal of putting a man into space before 1960 ended did not happen. The first Mercury flight, unmanned, did not take place until December 19, 1960. The second Mercury mission on January 31, 1961, carried as its passenger a 3-year-old chimpanzee named Ham. The flight lasted 18 minutes; Ham was weightless for 6.5 minutes.

Navy Cmdr. Alan B. Shepard Jr. awaits liftoff in the Mercury capsule *Freedom 7* on May 5, 1961. He had been rousted from bed at 1:05 a.m. in order to prepare for his history-making launch into space. Shepard had gone through this same routine on May 2 and May 4. Both launches had been scrubbed.

The 80-foot-high Mercury-Redstone rocket sits on Cape Canaveral's Pad 5 waiting for liftoff. Alan Shepard was on board the *Freedom 7* capsule. "Ignition. Mainstage. Lift-off." At 9:31 a.m., America's first astronaut was headed into space. Shepard's 15-minute, 22-second flight achieved a height of 116.5 miles above the Earth and traveled 302 miles downrange from the launch site.

Alan Shepard in *Freedom 7* splashed down in the Atlantic Ocean, east of the Bahamas, and was picked up by the aircraft carrier *Lake Champlain*. Three days later, Shepard was treated to a hero's welcome in Washington, D.C., where he was presented with the Distinguished Service Medal by President John F. Kennedy. The success of the mission was tempered only by the fact that the Soviet Union had launched cosmonaut Yury Gagarin into space a month earlier.

Mercury 4 astronaut Virgil "Gus" Grissom is suited up and ready to climb into the 10-by-6-foot *Liberty Bell 7* capsule on July 19, 1961. Unfortunately for Grissom, that day's launch of *Mercury 4* was scrubbed because of bad weather. With America's second spaceman inside *Liberty Bell 7*, the Mercury-Redstone lifted off the launch pad at Cape Canaveral at 6:00 a.m. two days later. *Liberty Bell 7* reached an altitude of 118 miles during Grissom's 15-minute-37-second flight.

After splashdown, Grissom prepared for extraction from *Liberty Bell 7*. The capsule was powered down; the hatch, armed with explosives, was ready to be opened. With a line attached to the capsule, a Marine helicopter was ready to retrieve it. Suddenly, the hatch blew open prematurely and water streamed into the capsule, forcing Grissom to evacuate. *Liberty Bell 7* sank; not until the summer of 1999 did the efforts to recover it prove successful.

John Glenn, the best known of all Mercury astronauts, climbed into *Friendship 7* before his launch on February 20, 1962. His aviation experience was unparalleled. A highly decorated World War II and Korean War hero, Glenn had shot down three MiGs in Korea. The Marine pilot earned the distinction of being the first American astronaut to undertake a space flight that would orbit the Earth.

John Glenn's capsule *Friendship 7* sat high atop the 95-foot-high Atlas booster on the morning of February 20. Glenn's trip into space seemed as if it would never come. *Mercury 6* had seven scrubbed launch dates before the magic day of February 20 when Glenn was launched into space at 9:47 a.m.

By the time John Glenn splashed down into the Atlantic Ocean 4 hours and 55 minutes following launch, he had orbited the Earth three times. Glenn received the Distinguished Service Medal and a ticker-tape parade in New York City. President John F. Kennedy, a strong proponent of America's space program, presented John Glenn with the NASA Distinguished Service Medal on February 23. Of his history-making flight Glenn said, "It is hard to beat a day in which you are permitted the luxury of seeing four sunsets."

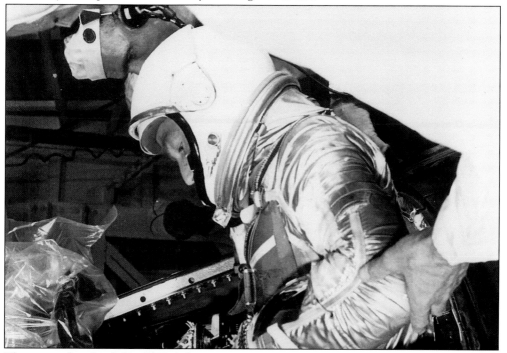

Three months after John Glenn's historic venture into space, M. Scott Carpenter set out to repeat the feat. As with almost every other Mercury mission, *Mercury 7* experienced several delays before successful liftoff on May 24, 1962. Following his 4-hour and 56-minute flight in which he orbited the Earth three times, Carpenter splashed down in the Atlantic Ocean.

A comedy of errors, including wasted fuel and a delay in firing the capsule's rockets, forced Scott Carpenter to splash down 250 miles off target. Because of a communications equipment problem, mission control did not know if Carpenter had splashed down, or worse yet, even if he had survived the reentry. *Aurora 7* floated in the Atlantic Ocean for three hours until the arrival of the recovery team.

The next Mercury astronaut to venture into space was Walter M. Schirra Jr. In his silver Mercury pressure suit, Schirra posed for his official NASA portrait in front of a model of the Mercury spacecraft. A war hero, Schirra had flown 90 combat missions in Korea, where he had downed a pair of Korean MiG 15s in air-to-air combat.

Carrying Walter Schirra in his *Sigma 7* capsule, the 100-foot Atlas rocket lifted off Kennedy Space Center's Pad 14 at 8:15 a.m. on October 3, 1962. Originally scheduled for a September launch, the mission had to be postponed because of a fuel leak in the Atlas booster. Barely five minutes after launch, the booster shut down, placing the *Sigma 7* space capsule into Earth orbit.

Mercury 8 astronaut Walter Schirra is pictured leaving his space capsule *Sigma 7* after its Pacific Ocean recovery. Schirra's flight included five-and-three-quarters Earth orbits and could not have gone better. As the *Sigma 7* capsule began reentry, the retrorockets fired perfectly. The drogue parachute opened right on schedule; the main parachute opened only minutes later. Splashdown was perfect; the landing bag deployed without a hitch.

At 4:55 a.m. on May 15, 1963, *Mercury 9* astronaut L. Gordon Cooper Jr. left his Hangar S headquarters headed for Cape Canaveral's Pad 14. Nicknamed Gordo, the youngest of the original Mercury astronauts had been assigned the most aggressive space flight to date; Cooper would log 22 Earth orbits while flying in space 34 hours. When an inoperable automatic control system required Cooper to land the spacecraft manually, he splashed down only four miles from the carrier *Kearsarge*.

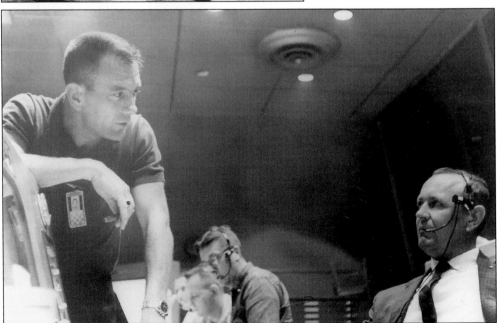

Prior to the launch of *Mercury 9*, piloted by Gordon Cooper, Donald K. Slayton and flight director Christopher C. Kraft discuss the mission. Slayton was the only Mercury astronaut who did not fly a Mercury mission. Scheduled to fly *Mercury 7*, he was removed from flight status only ten weeks before the scheduled launch date of May 1962 because of a preexisting heart murmur. Ten years later, Slayton regained his astronaut flight status when a NASA review board found him fit to fly.

Two

GEMINI
THE HEAVENLY TWINS

Putting a man into space for several hours as the Mercury program did was one thing. Sending two men into space for several days was an entirely different matter. That was what NASA envisioned when plans for a two-man program were announced on December 7, 1961. Designated as Gemini on January 3, 1962, the program was named after the third constellation of the zodiac, the twin stars Castor and Pollux. NASA intended Gemini as the means to prove that the United States could safely send a man to the Moon. The Soviet Union still held the lead in the space race, but not for long. Between March 23, 1965, and November 11, 1966, the United States launched ten successful manned Gemini missions. The Soviet Union had not launched

even one. Gemini proved America could go to the Moon, man could function outside of the space capsule, precision rendezvous and docking procedures were possible, and that man could endure extended periods of weightlessness. Because of the Gemini program, the United States had taken the lead in the space race and was almost ready for the journey to the Moon.

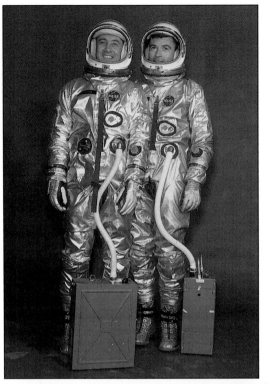

The first American two-man mission crew consisted of *Gemini 3* astronauts Virgil I. Grissom, commander, and John W. Young, pilot. The primary objectives of the mission included evaluation of the new two-man design, plus demonstrations of various systems and procedures. Secondary objectives called for testing of flight crew equipment, biomedical instrumentation, and the personal hygiene system. The astronauts were also expected to perform three scientific experiments while at the same time documenting their flight photographically.

After several delays resulting from the unmanned *Gemini 2* mission, *Gemini 3* was finally scheduled for launch on March 23, 1965. Liftoff was postponed for 24 minutes because of a leaking fuel line. Finally, with Gus Grissom and John Young aboard the capsule *Molly Brown*, *Gemini* lifted off Pad 19 at 9:24 a.m. Because of the *Gemini* capsule's doubled weight, the Atlas rocket used for Mercury launches had been replaced by a Titan intercontinental ballistic missile with a two-stage booster.

Following three Earth orbits, the *Gemini 3* mission ended after 4 hours and 52 minutes. Gus Grissom and John Young splashed down in the Atlantic Ocean only 58 miles from their target. While Grissom and Young remained in *Molly Brown*, navy frogmen from the USS *Intrepid* attached a flotation collar to the capsule as a helicopter hovered in the background.

Gemini 4 prime crew, commander James A. McDivitt and pilot Edward H. White, head for Pad 19 on the morning of June 3, 1965. The mission called for a four-day flight and the first space walk by an American astronaut. As air force majors, test pilots, and group 2 astronauts, McDivitt and White had a great deal in common.

Following a 76-minute hold because of problems with the launch pad erector, *Gemini 4* blasted into space at 10:16 a.m. on June 3, 1965. With an explosion of orange flames, the Titan 2 rocket lifted off the pad without a hitch. Within five minutes, *Gemini 4* was in Earth orbit at an altitude of 160 miles. As Edward White prepared to walk in space, *Gemini 4* sped through space at 17,500 miles per hour.

"Man does not live by bread alone." Nutritional meals were nearly as important to *Gemini 4* astronauts as the safety of their space capsule. Freeze-dried and dehydrated food such as beef and gravy, peaches, and strawberry cereal cubes provided four meals a day for James McDivitt and Edward White. The meals, stacked by meal by day, were stored in a compartment. When it was time to eat, the meals were reconstituted with water.

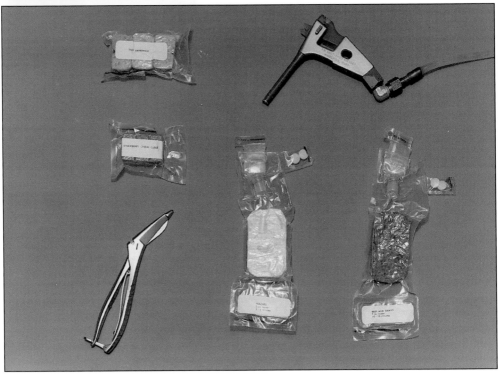

Gemini 4 pilot Edward White floats in the zero gravity of space. Originally scheduled for the mission's second orbit, preparations for the space walk took longer than planned, requiring a delay until the third orbit. White, attached to the capsule by a 25-foot umbilical line and a 23-foot tether line, walked in space for 21 minutes. Both lines were wrapped in gold tape, forming one cord to insure that they didn't become tangled. In his right hand, White carried a self-maneuvering unit that allowed him to steer in the direction he wished to go. Of his walk he recalled, "I was taking some big steps, the first on Hawaii, then California, Texas—lightly, in deference to the President—Florida, and the last on the Bahamas and Bermuda." Finally, an American had walked in space. As part of the *Voskhod 2* mission, the Soviets had already done it three months earlier.

The *Gemini 4* mission ended after four days and 62 Earth orbits with a splashdown in the Atlantic Ocean that was only 49 miles off target. Pictured, the *Gemini 4* capsule is being hoisted aboard the recovery ship, the USS *Wasp*. As the unshaven Edward White and James McDivitt exited the capsule, a U.S. Navy band played and sailors cheered. President Lyndon Johnson soon called with his congratulations.

Gemini 5 prime crew members L. Gordon Cooper Jr., mission commander, and Charles Conrad Jr., pilot, stand at the launch pad only days before their launch date of August 21, 1965. Objectives of the scheduled eight-day mission included evaluations of the rendezvous guidance and navigation system and of the effects of weightlessness on the crew.

Records come and go; so it was and is with space flight. Despite problems with oxygen pressure to the fuel cells, *Gemini 5* eclipsed the flight time of both *Gemini 4* and that of the Soviet *Vostok V5* mission. After seven days and almost 23 hours in space, *Gemini 5* splashed down in the Pacific Ocean. Charles Conrad and Gordon Cooper are shown on the deck of the USS *Lake Champlain* following recovery on August 29, 1965.

Suited up, *Gemini 7* pilot James A. Lovell Jr. and mission commander Frank Borman head for the launch pad on the morning of December 4, 1965. A 14-day mission, *Gemini 7* was designed to evaluate the long-term effects of space flight and weightlessness; a rendezvous with *Gemini 6* was also planned. The flight had faced postponement because of the delayed *Gemini 6* mission, but stayed in the schedule and lifted off at 2:30 p.m. on its originally targeted date. *Gemini 6* tried again days later.

A fish-eye camera lens provides a view of the forward displays and controls of the *Gemini 7* spacecraft. One of the most important objectives of this particular mission, the longest to date, was to study the effects of an extended space flight upon the crew. During the nearly 14-day flight, Frank Borman and James Lovell experienced almost constant weightlessness. Other difficulties faced by the astronauts included cramped quarters and work areas, a lack of privacy, and a steady diet of reconstituted meals. Such situations would have to be endured, however, if man were ever to travel to the Moon or beyond. The success of *Gemini 7* proved that long-term space flight was not just a concept but rather a reality; reaching the Moon was becoming increasingly possible.

Gemini 6 astronauts Thomas P. Stafford and Walter M. Schirra Jr. sit in their capsule on December 15, 1965, in preparation for launch. Soon the capsule's hatches would be closed and the countdown would begin yet again. Scheduled for an October launch, *Gemini* 6 was supposed to rendezvous with an Agena target. Set for liftoff less than two hours after an Atlas rocket took the Agena into space, *Gemini* 6 was scrubbed when the Agena blew apart six minutes into flight.

Gemini 6 seemed doomed; there had been another abortive liftoff attempt on December 12, 1965. Launch was rescheduled for three days later. Finally, everything was a go; Walter Schirra and Thomas Stafford were launched into space at 8:37 a.m. on December 15, while *Gemini* 7 was in orbit over Cape Canaveral. Only minutes after launch, *Gemini* 6 reached an altitude of 168 miles and entered orbit.

A familiar face is always a welcome sight when one is far from home. Less than six hours after launch, the *Gemini 6* astronauts caught a glimpse of their fellow Gemini astronauts. A distance of only 30 miles separated them from *Gemini 7*. Soon the distance closed to 120 feet. As Walter Schirra brought his craft to within one foot of *Gemini 7*, the moment of rendezvous was at hand. As *Gemini 6* and *Gemini 7* sped through space at a speed of 17,500 miles per hour, the capsules stayed in formation for 5 hours and 15 minutes. When this photograph of *Gemini 7* and the Earth below was taken from *Gemini 6* at a distance of only 37 feet, the historic meeting was documented for posterity. Never before had two manned American spacecraft met in space.

James Lovell is hoisted to a recovery helicopter while Frank Borman waits his turn. Lovell had piloted their *Gemini 7* capsule to an on-target splashdown on December 18, 1965, only 6.4 nautical miles from the USS *Wasp*. Recovery of the astronauts and their spacecraft was completed in near record time. The dye in the water surrounding the raft had been released to deter sharks and to facilitate spotting by the recovery team.

Compared to other Gemini and later Apollo missions, the flight of *Gemini 6* had been brief; it lasted slightly less than 26 hours. Although short in duration, all of the mission's objectives were realized. Rendezvous between vehicles in space was possible. After 16 orbits, *Gemini 6* splashed down on December 16, 1965, two days prior to the return of *Gemini 7*. Naval personnel welcomed Walter Schirra and Thomas Stafford onto the USS *Wasp* only 66 minutes later.

The relatively inexperienced *Gemini* 8 crew members David R. Scott and Neil A. Armstrong logged several firsts with their mission scheduled for March 15, 1966. Scott was the first group 3 astronaut to enter space; Armstrong was the first civilian astronaut launched into space. Prior to the *Gemini* 8 mission, Scott participated in a maintenance and repair experiment during a weightlessness exercise aboard a KC-135 aircraft.

The primary mission of *Gemini* 8, a rendezvous and hard dock, was accomplished when *Gemini* 8's cone eased inside the docking collar of this Agena target. Success was quickly followed by problems. An electrical short in *Gemini*'s control systems caused a violent whirling of the spacecraft that could be controlled only by use of the reentry control thrusters. Fearing for the astronauts' safety, NASA ordered a forced early separation, cancellation of Edward White's space walk, and *Gemini* 8's immediate return to Earth.

Gemini 9 astronauts Thomas P. Stafford, command pilot, and Eugene A. Cernan, pilot, are suited up at Pad 19 on the morning of June 3, 1966. Cernan and Stafford, the original Gemini 9 backup crew, became the prime crew when Elliott See and Charles Bassett died in an aircraft accident on February 26. Their T-38 crashed into the McDonnell facility in St. Louis, Missouri.

Plagued by constant problems, Gemini 9 was finally launched on the morning of June 3, 1966. The original launch scheduled for May 17 had been scrubbed because of the destruction of the Agena docking target on launch. A substitute Agena was successfully launched on June 1. Gemini experienced problems that day with its guidance-control computer. The scheduled three-day mission included docking with the Agena target and a walk in space by Eugene Cernan.

Gemini 9 entered orbit 185 miles above the Earth. Slightly more than four hours following launch, as *Gemini* 9 sped through its third orbit, Thomas Stafford and Eugene Cernan closed to within a mile of the Agena. Moving ever closer, the astronauts noticed a problem with Agena. Plans to dock were eventually scrapped. Just under 50 hours into the mission, Cernan climbed out into the darkness of space. His walk lasted two hours and seven minutes. Cernan took this photograph of the nose of *Gemini* 9 while he stood in the hatch of the capsule. Three days after launch, the mission ended with splashdown. *Gemini* landed less than 2 miles from the USS *Wasp*. Only an hour later, the crew was piped aboard the deck of the carrier. They received a congratulatory call from President Lyndon Johnson and began their rigorous medical and mission debriefing.

Gemini 10 crew members Michael Collins and John W. Young suit up in the trailer before their launch on July 18, 1966. The mission was billed as the "most ambitious manned space flight ever attempted by the United States." Mission objectives called for a rendezvous and docking with two different Agena rockets, 14 space experiments, and a walk in space by Michael Collins.

Once in orbit, calculation errors pushed *Gemini 8* off course. Corrections required the crew to burn more fuel than planned. By the fourth orbit, only six hours after launch, the first docking took place, as pictured. Still docked, the mated *Gemini* and Agena vehicles roared to a record altitude of 474 miles. After being docked together for almost 39 hours, *Gemini 10* separated from the Agena on July 20.

The *Gemini 11* prime crew members were Charles Conrad Jr., command pilot, and Richard F. Gordon Jr., pilot. The mission objective called for a rendezvous and docking with an Agena target, followed by a space walk by Richard Gordon. Conrad was a veteran of *Gemini 5*, but this was the first mission for group 3 astronaut Gordon.

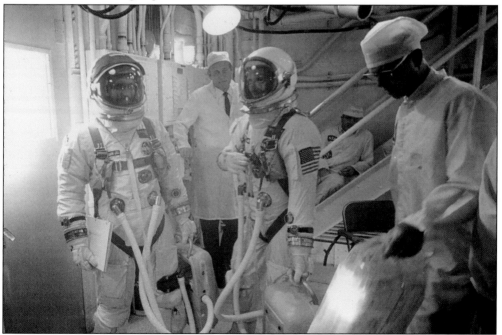

Suited up and ready to go, Richard Gordon and Charles Conrad prepare to enter the *Gemini 11* spacecraft in the White Room atop Pad 19 on September 12, 1966. The launch had already been postponed twice. On September 9, the mission was scrubbed because of a leak in the first stage oxidizer; the next day, the mission was again canceled because of a problem with the autopilot.

On September 12, 1966, Charles Conrad and Richard Gordon for the third time in only four days went through the elaborate prelaunch ritual. The Agena target had already been launched at 8:05 a.m. Early inspection found a suspected leak around the command pilot's hatch. Countdown was interrupted with a 16-minute hold. Finally at 9:42 a.m., Pad 19 shook, red flames erupted from the Titan rocket, and *Gemini 11* lifted off.

Ninety-four minutes after launch, the Agena was in sight. The two vehicles were quickly docked. The next day, leaving *Gemini*, Richard Gordon half walked, half floated to the Agena. He carried a line with him to tether the two craft. Gordon climbed onto the nose of the *Gemini* capsule, as if riding a horse. From the capsule Charles Conrad yelled, "Ride 'em cowboy!" Pictured is Gordon returning to *Gemini*.

Because Richard Gordon's space suit had been having a problem with overheating, the walk was terminated early. At midnight on the second day of the mission, the docked spacecraft soared to a record height of 850 miles. After decreasing altitude, the hatch was again opened and Gordon hung out of the capsule to snap hundreds of photographs. Fifty hours into the mission, *Gemini* and Agena separated. Following a second rendezvous with Agena, *Gemini 11* returned to Earth.

Gemini 12 astronauts James A. Lovell Jr., mission commander, and Edwin E. "Buzz" Aldrin, pilot, check their cameras only hours before their scheduled launch. The final Gemini mission was an ambitious one. NASA mission control had planned a rendezvous and docking, 14 experiments, a tethered vehicle operation, a space walk of several hours, and a test of the automatic reentry system.

On November 9, 1966, technicians suspected an autopilot problem with the Atlas booster, which led to a two-day delay. Launch was now scheduled for November 11, Veterans Day. In the White Room, James Lovell and Edwin Aldrin were treated to a bit of humor. A poster on the wall read, "Last chance. No relaunch. Show will close after this performance." *Gemini 12* lifted off at 3:46 p.m., as scheduled.

Gemini 12 entered Earth orbit only six minutes following liftoff. The rendezvous with the Agena target was planned for the mission's third orbit. The rendezvous was complete 3 hours and 46 minutes after launch. Docking took place 28 minutes later. Less than 20 hours into the mission, Edwin Aldrin walked in space for 2 hours and 29 minutes and took this photograph of the *Gemini 12* capsule.

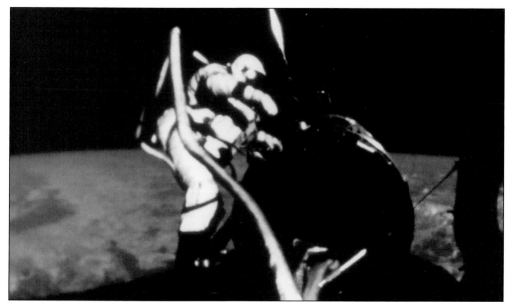

The second extravehicular activity (EVA) of *Gemini 12* took place at almost 43 hours into the mission. Edwin Aldrin used a mounted handrail on the side of the spacecraft to slide along the capsule to the nose. He spent 2 hours and 8 minutes outside the craft. Separation from Agena took place at 52 hours and 14 minutes. Aldrin participated in yet one more EVA when he stood in the capsule's open hatch for 55 minutes.

James Lovell and Edwin Aldrin receive a hero's welcome as they return to Kennedy Space Center on November 16. Thousands of NASA, military, and contractor personnel turned out to greet the pair as the Gemini program came to an end. Each astronaut claimed personal records. Aldrin broke the record for extravehicular activity; Lovell now had more time in space than any other man.

The final mission of the Gemini program had been an unqualified success. *Gemini 12* splashed down in the Atlantic Ocean within 3 miles of the USS *Wasp* on November 15, 1966. James Lovell and Edwin Aldrin had logged 3 days, 22 hours, and 34 minutes in space; total time for extravehicular activity had been 5.5 hours. *Gemini 12* had completed 59 Earth orbits; its crew had performed 14 experiments during the mission. Following their return to Kennedy Space Center, Lovell and Aldrin shook hands in front of a scoreboard sign that listed all of the manned Gemini flights. An era had ended. Challenges had been met and obstacles conquered. Other historic journeys into space were still to come. The Moon was next.

Three

THE APOLLO PROGRAM

On May 25, 1961, President John F. Kennedy called for a space mission to take an American to the Moon. To say that it was an overly aggressive call to arms would be a tremendous understatement. The first American-manned flight, *Mercury 3,* which lasted only a few minutes, had been flown only weeks before. The technology just did not exist. The goal of NASA and the President had become one of putting an American into space, landing him on the Moon, and returning him safely to the Earth before the Soviets could do it. To accomplish this, NASA needed "to establish the technology to meet other national interests in space, to achieve preeminence in space for

the United States, to carry out a program of scientific exploration of the Moon, and to develop man's capability to work in the lunar environment." The objective was clear; America was going to the Moon. Although the Mercury and Gemini projects had laid the groundwork, it would take more than lofty words from a young and charismatic president. Along the way, there would be tremendous victories and defeats; men would die to make it happen.

The *Apollo 1* prime crew members were Virgil I. Grissom, Edward H. White II, and Roger B. Chaffee. *Apollo 1*, intended as a shakedown flight for the Apollo program, was originally scheduled for a December 1966 launch. That date came and went, as the mission was rescheduled for February 1967. The two-week mission seemed destined for failure from the beginning. There were constant problems with numerous components during ground testing.

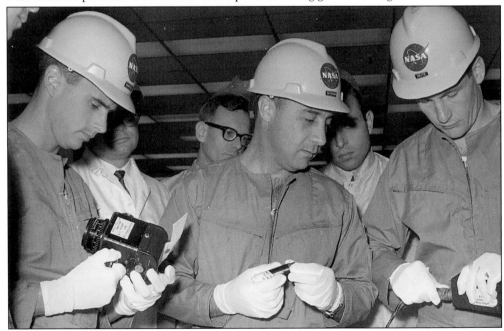

In the foreground *Apollo 1* astronauts, from left to right, Roger B. Chaffee, Virgil I. Grissom, and Edward H. White II inspect command module equipment at North American Rockwell's headquarters in Downey, California. North American engineers and technicians stand in the background. North American, the prime contractor for the Apollo command module, and NASA shared the same goal: construction of safe and sound operational equipment.

Apollo 1 was being ground tested at Complex 34, Cape Kennedy, Florida, on January 27, 1967. Crew members had been working in the capsule for nearly six hours when an electrical short occurred. At 6:30 p.m. a flash fire broke out in the spacecraft. Roger Chaffee screamed, "We've got a bad fire . . . We're burning up here!" Sixteen seconds later, the intense heat had destroyed the command module interior, killing all three astronauts. This photograph of the charred exterior of the *Apollo 1* command module showed the effects of the intense heat from the fire. While other astronauts had been killed in plane crashes or car accidents while enrolled in the space program, this was the first time that deaths had occurred during the course of an actual NASA mission. The tragedy set America's Moon program back by at least two years.

Donn F. Eisele, Walter M. Schirra Jr., and Walter Cunningham were selected to serve as the prime crew for *Apollo 7*, the first manned Apollo mission. The 45-year-old Schirra, an original Mercury astronaut, was highly experienced, having flown on both the one-man *Mercury* 8 and two-man *Gemini* 6 missions. Both Cunningham and Eisele were group 3 astronauts who had yet to experience actual space flight.

The first stage of the Saturn 1B launch vehicle sits at Kennedy Space Center's Pad 34. Due to the *Apollo 1* tragedy, the *Apollo 7* mission experienced even closer scrutiny than normal. The equipment had undergone substantial modifications. After a delay of 19 months, *Apollo 7* was expected to put the Apollo program back on track.

The Saturn 1B rocket carrying *Apollo 7* astronauts Walter Schirra, Walter Cunningham, and Donn Eisele was launched from Kennedy Space Center's Complex 34 at 11:03 a.m. on October 11, 1968. Understandably, mission control monitored the mission with great intensity. Mission control had little cause for concern; the 10-day, 20-hour *Apollo 7* mission was remarkably free of problems. The mission was everything NASA had hoped it would be.

After 163 Earth orbits, 4.5 million miles flown, and almost 11 days in space, it was time for *Apollo 7* to return to Earth. *Apollo 7*'s splashdown went as smoothly as the flight. Walter Schirra, Walter Cunningham, and Donn Eisele splashed down at 7:11 a.m. on October 22, 1968, approximately 200 nautical miles south-southwest of Bermuda. Assisted by a team of navy frogmen, Schirra climbed out of the capsule.

Apollo 7 crew members Walter Schirra, Donn Eisele, and Walter Cunningham arrive aboard the USS *Essex*, following their Atlantic pickup. They promptly received a congratulatory telephone call from President Lyndon B. Johnson. Apollo program director Sam Phillips declared the *Apollo 7* mission "101 percent successful." America's space program was back on track. America was on its way to the Moon.

With their *Apollo 8* mission scheduled for December 21–27, 1968, command module pilot James A. Lovell Jr., lunar module pilot William A. Anders, and mission commander Frank Borman were going to spend Christmas far from home. Instead of traditional festive dinners back on Earth, their holiday fare was to be freeze-dried rehydrated meals, as the crew entered the history books as the first Americans to orbit the Moon.

Apollo 8's Saturn V rocket was attached to the mobile launch platform at Kennedy Space Center on December 21, 1968, the first time the Saturn V launch vehicle had been used for a manned flight. The Apollo 8 crew waited in anticipation of the 7:51 a.m. launch. Liftoff was flawless; the Saturn V operated just as NASA's mission control hoped.

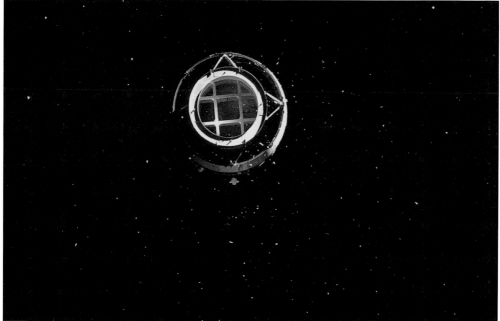

The third stage of the Saturn V had just separated from the spacecraft, as pictured in this photograph taken from the Apollo 8 spacecraft. The three stages of the 363-foot-tall Saturn V rocket produced millions of pounds of thrust generated by 15 engines. The first manned mission to circle the Moon, Apollo 8 logged 10 lunar orbits that took 20 hours.

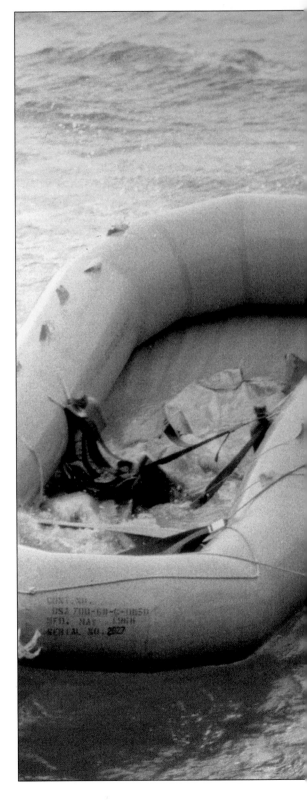

On Christmas Eve, the astronauts had ended a live television broadcast from space by reading a passage from Genesis regarding the formation of the heavens and Earth. They then ended their message by saying "and from the crew of *Apollo* 8, we close with good night, good luck, a Merry Christmas, and God bless all of you—all of you on the good Earth." The next morning the crew started for home, ending their six-day, three-hour mission with splashdown on December 27, 1968, only 3 miles from the USS *Yorktown*. Within minutes, a helicopter hovered over the *Apollo 8* capsule. United States Navy frogmen quickly attached a flotation collar to the bobbing spacecraft. Forty minutes later James Lovell and then William Anders and finally Frank Borman emerged from the capsule and climbed into a life raft. After spending Christmas orbiting the Moon, the astronauts were ready for a post-holiday reunion with families.

Apollo 8 had been a tremendous accomplishment. New ground had been paved and new equipment had been tested with great success. All mission objectives had been achieved. Thousands of photographs, such as this one of the Moon's surface, had been taken. Lovell described the Moon as "essentially gray, no color. Looks like plaster of Paris. Sort of grayish sand."

David R. Scott, James A. McDivitt, and Russell L. Schweickart were the crew members for *Apollo* 9. While McDivitt and Scott were both veterans with Gemini experience, this was to be Schweickart's first and only venture into space. McDivitt and Schweickart are shown working inside of the primary Apollo lunar module simulator at Kennedy Space Center in preparation for the *Apollo* 9 mission.

Countdown for *Apollo 9* was started on February 22, 1969. Liftoff for the mission was scheduled six days later. When all three crew members developed a mild respiratory illness, the mission was delayed to allow for their recovery. James McDivitt, David Scott, and Russell Schweickart sat in the *Apollo 9* command module named *Gumdrop* on the morning of March 3. Precisely at 11:00 a.m. the powerful Saturn rocket lifted off Pad 39.

The launch was so smooth that *Apollo 9* commander James McDivitt called it "an old lady's ride." When stage one's five engines stopped, the crew lurched forward. As stage two cut in, the result was just the opposite: the three men were violently pushed backward. Slightly over 11 minutes into the mission, the Saturn's third stage put *Apollo 9* into orbit. Pictured are the docked command and service modules and the lunar module.

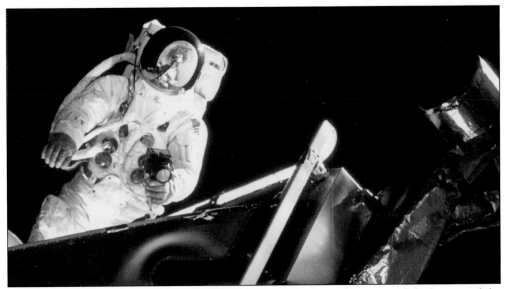

On day three of the mission, James McDivitt and Russell Schweickart entered the lunar module. After firing the lunar module's descent-propulsion system, they returned to the command module. The next day they were back in the lunar module. Schweickart performed a 37-minute extravehicular activity, after which they returned to the command module. Walking between the open hatches of the lunar and command modules, Schweickart used a Hasselblad camera to take hundreds of photographs.

The *Apollo 9* lunar module *Spider* was photographed from the command module. On March 7, 1969, James McDivitt and Russell Schweickart crawled through the hatch of the lunar module. For the next 6.5 hours, the modules flew separately at distances of up to 110 miles. After docking again, McDivitt and Schweickart reunited with David Scott in the command module. Ten days after launch, *Apollo 9* splashed down in the Atlantic Ocean.

Apollo 10 astronauts Eugene A. Cernan, Thomas P. Stafford, and John W. Young pose for an official photograph in front of a mockup of the Moon's surface. Scheduled for mid-1969, the *Apollo 10* mission was to be a practice run for the anticipated manned lunar landing—the final test of both the command and lunar modules.

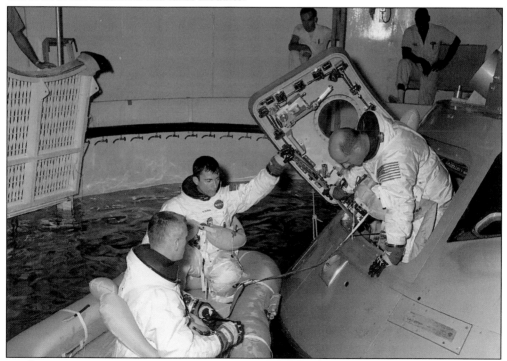

Apollo 10 astronauts take part in a water-egress training exercise at the Manned Spacecraft Center. Since American space missions ended with ocean splashdowns, every effort was made to ensure that astronauts were as familiar with the procedure as possible. Astronauts had to be well-trained in the event that something went wrong at splashdown.

Orange flames lick the first stage of the Saturn V rocket, as the crew of *Apollo 10* is launched from Kennedy Space Center's Pad 39B at 12:49 p.m. on May 18, 1969. Liftoff went without a hitch. Within seconds, the three-stage rocket pierced a high cloud base and disappeared from sight. Only minutes later, *Apollo 10* was at an altitude of 116 miles above the Earth.

The *Apollo 10* command module *Charlie Brown* is visible from *Snoopy*, the lunar module. When photographed, *Charlie Brown* was approximately 175 miles above the Moon. The lunar module, in a simulation of the upcoming *Apollo 11* manned lunar landing, was flown to a distance of only 9.4 miles from the Moon. The eight-day, three-minute flight of *Apollo 10* had been an unqualified success. America was ready for the Moon.

Four

To the Moon

American astronaut Neil A. Armstrong, commander; Michael Collins, command module pilot; and Edwin E. Aldrin Jr., lunar module pilot, were going to do what no one had ever done—they were going to the Moon. All three had previously flown in space as part of the Gemini project, respectively *Gemini 8, Gemini 10,* and *Gemini 12.* The 20-manned Mercury, Gemini, and Apollo flights that had gone before had all been preparation for the mission of *Apollo 11*—a manned lunar landing and a walk on the surface of the Moon. The father of America's space program, German-born Wernher von Braun, had described the Saturn V rocket as a collection of "thousands of parts and all built by the lowest bidder." Now it would carry three Americans to the Moon.

Apollo 11 astronauts Neil Armstrong, Michael Collins, and Edwin Aldrin prepare to enter the van that will take them to Kennedy Space Center's Pad 39A on the morning of July 16, 1969. They followed the same routine as had 35 American astronauts before them. They awakened early, ate a hearty breakfast, suited up, and were driven to the vehicle that would take them into space.

Apollo 11's command module *Columbia* and the lunar module *Eagle* sit atop the 363-foot-tall Saturn V rocket. The crew had been strapped in and connected to the aircraft's life-support system. At breakfast that morning, NASA's administrator Dr. Thomas Paine emphasized the safety of the astronauts as paramount. Should the crew be forced to abort the mission, they would be assigned another flight as quickly as possible.

A billion people worldwide watched their television sets, as orange flames and heavy white smoke engulfed Pad 39 when *Apollo 11* lifted off at 9:37 a.m. on July 16, 1969. Hundreds of thousands of people on Florida's beaches watched the three-stage Saturn rocket as it blasted upward. Only a minute after launch, Neil Armstrong, Edwin Aldrin, and Michael Collins were being hurled forward faster than the speed of sound. *Apollo 11* was in Earth orbit ten minutes later.

On their way to the Moon, *Apollo 11* astronauts photographed the Earth at an altitude of 10,000 nautical miles. The first two stages had separated perfectly. The third stage had taken *Apollo 11* to an orbital speed of 17,400 miles per hour. Neil Armstrong, mission commander, told Houston, "This Saturn gave us a magnificent ride. We have no complaints with any of the three stages on that ride. It was beautiful."

Lunar module pilot Edwin Aldrin is visible inside the lunar module *Eagle*, as it descends toward the Moon on July 20, 1969. As Aldrin and Neil Armstrong looked forward to accomplishing their history-making first manned lunar visit, Michael Collins, pilot of the command module *Columbia*, was left alone.

As the *Apollo 11* lunar module *Eagle* descended toward the Moon, Edwin Aldrin, the lunar module pilot, voiced the readout sequence to Armstrong as it was transmitted to Houston. "540 feet . . . 400 feet . . . 30 feet . . . drifting to right a little . . . OK." Finally, from Houston, "We copy you down, *Eagle*." Seconds later, Neil Armstrong spoke to Houston: "Houston, Tranquillity Base here. The *Eagle* has landed." Man was on the Moon. America had done it.

Less than seven hours after *Eagle* had landed, it was time to set foot on another world—the Moon. As Neil Armstrong stepped onto the surface of the Moon, he said, "That's one small step for man, one giant leap for mankind." Less than 15 minutes later, Armstrong photographed Edwin Aldrin as he backed down the steps of the lunar module. Two American astronauts were now standing on the surface of the Moon.

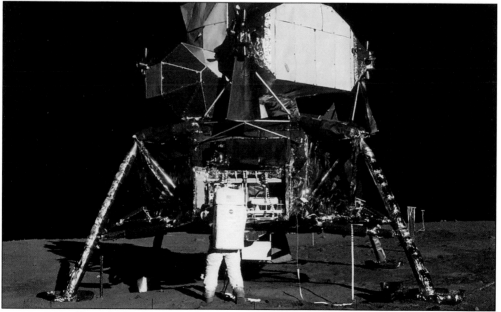

Using a 70-millimeter lunar surface camera, Neil Armstrong photographed the *Apollo 11* lunar module as it sat on the surface of the Moon on July 20, 1969. Buzz Aldrin was preparing to deploy the EASEP, or early Apollo scientific experiments package, that had been stowed in the lunar module's scientific experiment bay.

Edwin Aldrin, wearing a backpack that would keep him alive in the Moon's atmosphere for four hours, walks on the surface of the Moon near a leg of the lunar module *Eagle*. In their bulky space suits, the pair of *Apollo 11* astronauts moved as if in slow motion. On the Earth, garbed in the heavy suits and backpacks, each astronaut weighed approximately 360 pounds. On the surface of the Moon, they weighed only 60 pounds. Most of the photographs taken on

the Moon are of Edwin Aldrin. The reason was simple. Aldrin recalled, "Neil [Armstrong] had the camera most of the time." The astronauts' extravehicular activity on the Moon lasted 2 hours and 31 minutes. Several experiments were conducted, and 48 pounds of Moon rocks were collected.

American patriotism rang true, even on the Moon. The flag of the United States of America soon was flying on the Moon. Planting the flag had not been an easy process. Neil Armstrong and Edwin Aldrin had difficulty sticking the flagpole deep enough into the Moon's surface. Next, came a congratulatory telephone call from President Richard M. Nixon, who told the two *Apollo 11* astronauts, "Because of what you have done, the heavens have become a part of man's world."

After 21.6 hours on the Moon, it was time for *Eagle* to head for its rendezvous with *Columbia*, the command module. Less than four hours later, Michael Collins, Neil Armstrong, and Edwin Aldrin were reunited. They returned to Earth on July 24, splashing down in the Pacific Ocean. Immediately placed in a mobile quarantine facility aboard the USS *Hornet*, where effects of their lunar visit were studied, the astronauts received a congratulatory visit from President Nixon.

This spectacular view of the Earth rising over the Moon's horizon was taken from the *Apollo 11* spacecraft. On the last night before their return to Earth, each member of the crew voiced his thanks to everyone—the scientists, the American people and their Congressional representatives, and the industry and space agency teams—who had made the flight a reality. Edwin Aldrin summarized their sentiments when he said, "This has been far more than three men on a mission to the Moon; more, still, than the efforts of a government and industry team; more, even, than the efforts of one nation. We feel that this stands as a symbol of the insatiable curiosity of all mankind to explore the unknown. We've been pleased with the emblem of our flight, the eagle carrying an olive branch, bringing the universal symbol of peace from the planet Earth to the Moon." The 8-day, 3-hour, and 18-minute mission had been everything that President John F. Kennedy could have hoped for when he had called for such a mission only eight years earlier. NASA accomplished everything that it had set out to do with the *Apollo 11* project. America had gone to the Moon and returned safely. Armstrong and Aldrin had made ". . . one small step for man, one giant leap for mankind."

The prime crew of *Apollo 12* was comprised of Charles Conrad Jr., Richard F. Gordon, and Alan L. Bean. Conrad and Gordon were old hands at space flight. Conrad had been part of the *Gemini 5* and *Gemini 11* missions. Gordon had been a member of the *Gemini 11* mission. Scheduled for November 14, 1969, *Apollo 12* was to be Bean's first space mission.

Alan L. Bean is pictured outfitted in his white pressure suit. Although Bean was the rookie of the *Apollo 12* flight crew, he was extremely well-qualified. A naval officer, as were Richard Gordon and Charles Conrad, Bean was a graduate of the University of Texas and the Navy Test Pilot School, and had been a member of a fighter squadron based in Florida. He had joined the elite corps of astronauts in 1963.

Sunshine and thunderstorms had alternated the previous day; the day of the launch offered only a small window for liftoff. Postponement seemed inevitable, as bad weather threatened the launch. With a ceiling of only 800 feet and heavy thunderstorms, mission control decided on a go. The Saturn V rocket carrying *Apollo 12* into space lifted off the launch pad in a cloud of smoke on the morning of November 14, 1969.

When the Saturn was hit by lightning on launch, mission control considered aborting the mission, as the rocket sped through the clouds. Despite the difficulties, four days later, Charles Conrad and Alan Bean entered the lunar module. At 11:16 a.m. on November 18, 1969, the lunar module separated from the command module. The next day, America's second lunar landing took place. Bean climbed down the steps of *Intrepid* to join Conrad on the Moon.

Charles Conrad aligns the antenna on the lunar surface experiment package. The two astronauts planted an American flag and collected 50 pounds of Moon rocks. The mission's first extravehicular activity lasted almost four hours. The next day, Conrad and Alan Bean again walked on the Moon for nearly four hours. After 31.5 hours on the Moon, the lunar module blasted off to return to the command module. The mission ended with splashdown on November 24, 1969.

Fred W. Haise Jr., James A. Lovell Jr., and Thomas Mattingly Jr. made up the prime crew for the *Apollo 13* mission. The three are shown here standing at Kennedy Space Center's Pad 39A, site of their upcoming launch. Before the launch Mattingly was exposed to German measles and therefore was replaced by John L. Swigert Jr.

John Swigert, command module pilot, relaxes in the suiting room in anticipation of the *Apollo 13* launch on April 11, 1970. The tremendous success of the Apollo program had a downside. The American public had become lackadaisical about space flight. It had become too routine. Success had also led to budget cuts for the program. It had already been done. Why do it again?

Apollo 13 astronaut Fred W. Haise Jr. poses next to a model of the Moon. *Apollo 13*'s mission was a lunar landing in the area of Fra Mauro; it was to be the third manned lunar landing. Less than six minutes after a flawless launch on April 11, 1970, the crew felt an unusual vibration. An engine in the second stage shut down two minutes early, forcing remaining second-stage and third-stage engines to burn longer than programmed.

This view of the *Apollo 13* service module was photographed from the lunar module/command module. Shaking off the postlaunch glitches, the flight proceeded smoothly. Capsule communicator Joe Kerwin, at 46 hours and 43 minutes into the mission, told the *Apollo* crew members, "The spacecraft is in real good shape as far as we are concerned. We're bored to tears down here." Things were about to change. "OK, Houston, we've had a problem here," said John Swigert at 10:08 p.m. on April 13. Oxygen tank number two had blown up. Then, number one failed. The command module's normal supply of electricity, light, and water was inoperative. *Apollo 13* was 200,000 miles from home. Suffocation from lack of oxygen seemed a real possibility. Since the oxygen system in the lunar module was operational and ample for their emergency return to Earth, the trio quickly climbed into the lunar module. Power and water were both in short supply. It was up to *Aquarius*, the lunar module, to get the Moon-bound crew back to Earth.

Once the crippled capsule rounded the Moon, an engine burn sent it toward Earth. There were fears that the capsule was drastically off course. Another engine burn was required. If it failed, *Apollo 13* would be lost forever. When the burn ended, three words uttered by James Lovell told the story. "We've got it." They were on their way home.

Shortly before reentry, the crew returned to the command module. The lifesaving lunar module *Aquarius* was jettisoned. When it was all over and the three were safely on Earth, NASA administrator Thomas Paine said, "There has never been a happier moment in the United States space program. Although the *Apollo 13* mission must be recorded as a failure, there has never been a more prideful moment." The *Apollo 13* crew was welcomed aboard the USS *Iwo Jima* following splashdown on April 17, 1970.

"Astronauts Safe," read the headline of the *Honolulu Star-Bulletin* held by James Lovell. While praying for the well-being of the men, millions of Americans turned on their porch lights to provide a guiding beacon home from space. Across the globe, people of all nations now celebrated the return of the *Apollo 13* crew. Only days earlier, Lovell, now safely aboard the USS *Iwo Jima*, had feared the worst. Of the *Apollo 13* experience he recalled, "I looked out the window and saw this venting . . . my concern was increasing all the time. It went from 'I wonder

what this is going to do to the landing' to 'I wonder if we can get back home again . . .' and when I looked up and saw both oxygen pressures . . . one actually at zero and the other one going down, it dawned on me that we were in serious trouble." While American astronauts had died in car accidents, plane crashes, and the terrible *Apollo 1* fire, none had ever been lost in space. A near tragedy had been averted.

Alan B. Shepard Jr., Stuart A. Roosa, and Edgar D. Mitchell were the *Apollo 14* crew members. Only nine months earlier, the astronauts of *Apollo 13* had almost been lost in space during a death-defying flight. There was little excitement felt for the launch of *Apollo 14*. Few Americans cared about this mission. As scheduled, *Apollo 14* lifted off the launch pad on January 31, 1971.

The lunar module *Antares*, crewed by Alan Shepard and Edgar Mitchell, landed on the Moon on February 5, 1971. Ten years after his initial experience as the pilot on America's first manned spaceflight, Shepard was back in space. This time he walked on the surface of the Moon and planted the third American flag there. Mitchell soon followed Shepard to become the sixth man to walk in the gray lunar dust.

During the second extravehicular activity of *Apollo 14*, Edgar Mitchell studied a lunar map. This second walk on the Moon lasted slightly over 4.5 hours. The map was necessary; while headed to Cone Crater, Mitchell and Alan Shepard nearly became lost. After 33.5 hours on the Moon, *Antares* blasted off and returned to Stuart Roosa in the orbiting command module. *Apollo 14* returned to Earth on February 9, 1971, with splashdown in the South Pacific.

The *Apollo 15* crew members were David R. Scott, Alfred M. Worden, and James B. Irwin. Scott, mission commander, was an experienced space traveler, a veteran of both the Gemini and Apollo programs. Irwin and Worden were both about to experience the thrill of a launch for the first time. As veteran air force pilots, the crew had much in common.

Apollo 15 astronauts David Scott, Alfred Worden, and James Irwin lifted off Pad 39A at Kennedy Space Center at 9:34 a.m. on July 26, 1971. Although the weather was perfect that morning, the launch pad had been struck by lightning almost a dozen times during the previous month. President Richard M. Nixon called *Apollo 15* "the most ambitious exploration yet undertaken in space." Wearing improved space suits that provided increased mobility and allowed longer time outside of the lunar module, the astronauts performed three extravehicular activities for a total of 10 hours, 36 minutes. On the way back to Earth, Worden performed the first space walk outside of Earth's orbit—a 38-minute extravehicular activity to retrieve film from the side of the spacecraft. During the nearly 13-day mission, the crew spent 145 hours in lunar orbit and deployed a small subsatellite.

The hatch of *Apollo 15*'s lunar module *Falcon* opened at 9:13 a.m. on July 31, 1971. David Scott climbed out of the capsule and became the seventh man to touch the surface of the Moon. James Irwin, the eighth man on the Moon, soon joined him. Irwin worked on the battery-powered lunar rover vehicle, the 10-foot-long car that had a top speed of 9 miles per hour.

The lunar rover allowed the astronauts to travel more than 27 kilometers (17 miles) across the surface of the Moon. During one exploration, the men found the "Genesis Rock," a chunk of lunar crust that was extensively studied to learn more about the origins of the Earth and Moon. This mission was the first time the lunar rover had been used.

The *Apollo 16* prime crew was comprised of Thomas K. Mattingly II, John W. Young, and Charles M. Duke Jr. One month prior to their scheduled launch date, mission commander Young, Mattingly, and Duke briefed the employees of the Kennedy Space Center on their mission. The *Apollo 16* crew then went through the audience of 1,500 people shaking hands and signing autographs. Shortly afterward, the crew went into their three-week prelaunch quarantine.

Apollo 16 astronauts engage in an extravehicular activity (EVA) exercise in advance of their mission. Practice, practice, practice were the keywords in preparing for launch, mechanical problems, landings, and even EVAs. Partial gravity simulators allowed astronauts in training to "walk in space" in their space suits under conditions of virtual reality; by use of the simulator, their weight was only one-sixth of what it actually was.

During the first of three extravehicular activities, John Young salutes the American flag near the landing site of *Orion*, their lunar module. The successful April 16, 1972, launch almost ended in an aborted descent to the Moon. On April 20, just prior to its descent, the lunar module developed problems with its engine controls. The backup guidance system was switched on and the lunar landing was a go.

During the second walk on the Moon by John Young and Charles Duke, Young collected soil and rock samples. The 11-day *Apollo 16* mission was one of the longest, most ambitious, and probably the most productive of all manned lunar missions. In 71 hours on the Moon, the crew took part in three extravehicular activities, traveled a total of 12 miles in the lunar rover, and collected 210 pounds of rock and soil samples.

During their third walk on the Moon, John Young put the lunar rover through its paces in a "Grand Prix" race at the Descartes landing site. Charles Duke filmed the event with a handheld, 16-millimeter motion-picture camera. The lunar rover not only expanded the astronauts' range of exploration on the Moon's surface but also facilitated the collection of hundreds of pounds of Moon rocks.

The prime crew of *Apollo 17,* Eugene A. Cernan, Harrison H. Schmitt, and Ronald B. Evans, pose with the lunar rover during the rollout of the final Apollo mission at the Kennedy Space Center. Schmitt, the lunar module pilot, was to become the first astronaut-scientist to land on the Moon. He and Cernan were to walk on the Moon while Evans piloted the command module *America* in lunar orbit.

As the Saturn V space vehicle lifted off Pad 39A at 12:33 a.m. on December 7, 1972, the glow filled the night sky for miles around the Kennedy Space Center. The objective of *Apollo 17*, the final Apollo manned lunar mission, was a landing in the Taurus-Littrow highlands and valley area of the Moon. The area contained rocks both older and younger than those gathered on previous missions.

After a four-day journey, Eugene Cernan and Harrison Schmitt touched down on the Moon. On their lunar landing at 2:54 p.m. on December 11, Cernan said, "The *Challenger* has landed. We is here. Man, we is here!" Cernan was first out of the hatch. He became the 11th man on the Moon. Schmitt soon followed. Schmitt took this photograph of Cernan and the lunar roving vehicle.

During the final *Apollo 17* lunar walk, Eugene Cernan photographed Harrison Schmitt standing next to a huge boulder in the Moon's Taurus-Littrow valley. *Apollo 17*'s mission included three extended lunar extravehicular activities (EVAs) by Cernan and Schmitt, totaling 22 hours and 4 minutes, plus a trans-Earth EVA by Ronald Evans. Although the television cameras frequently provided the American public with glimpses of the space travelers during relaxing moments, most of the time during missions was spent working. Among the scientific objectives of this particular mission were the collection of materials and geological surveying of this particular area of the Moon's surface. Successfully completing this chore, Schmitt and Cernan collected a record 243 pounds of lunar samples. The astronauts also set up the sixth automated research station on the Moon, deploying experiments to record heat flow, seismic activity, surface gravity, and atmospheric composition. In addition, they conducted various in-flight experiments, completed photographic tasks, and performed biomedical experiments.

Apollo 17 commander Eugene Cernan salutes the final American flag to be deployed on the Moon. Partially obscured in the background is the battery-powered lunar rover vehicle (LRV). Used extensively on this mission, the LRV was driven 30.5 kilometers during the three extravehicular activities. Cernan and Schmitt spent a record 75 hours on the Moon. During the trip back to Earth, Ronald Evans, command module pilot, spent one hour and six minutes walking in space.

Apollo 17 returned to Earth on December 19, 1972. An era had ended. Although Apollo began with a tragedy and the loss of three American astronauts, the program ended with the very successful *Apollo 17* mission. During a four-year period, there were 11 lunar missions and six Moon landings. Twelve Americans walked on the lunar surface. The Moon had been conquered. America had won the space race.

Five

AMERICA'S ORBITAL SPACE STATION

If there is one fact regarding space travel that is universally recognized, it is that it is expensive. Early in the space race, both NASA administrators and government officials understood that a single-use space vehicle was not practical. A reusable spacecraft was needed. Planning had begun as early as the mid-1960s to produce an orbiting space station that would serve as a science and engineering laboratory. Originally called the Apollo Applications Program, the project became Skylab in the 1970s. Skylab's objectives were twofold: "to prove that humans could live and work in space for extended periods and to expand our knowledge of solar astronomy well beyond Earth-based observations."

Launched on May 14, 1973, America's first reusable space vehicle, *Skylab 1*, began a six-year program. During the course of three separate missions, three-man crews consisting of a commander, pilot, and scientist-pilot spent 171 days and 13 hours in orbit at an altitude of 270 miles above the Earth. Serving as a transition between the Apollo project and reusable shuttle missions, Skylab completely met all of NASA's objectives.

The unmanned *Skylab 1* workshop was launched from Kennedy Space Center on May 14, 1973. Sitting atop a Saturn V rocket, the 85-ton laboratory was placed in orbit at 270 miles above the Earth. One minute after liftoff, *Skylab*'s meteoroid shield unexpectedly opened and ripped from the space station. The usefulness of *Skylab* was immediately in question.

Scheduled for May 15, 1973, the launch of *Skylab 2* aboard a Saturn IB rocket and its 28-day rendezvous with *Skylab 1* were postponed because of the damage to the *Skylab* laboratory. *Skylab 2* crew members included Paul J. Weitz, pilot; Charles Conrad Jr., commander; and Joseph P. Kerwin, scientist-pilot.

During a prelaunch training session, Paul Weitz suits up in Building 5 at Johnson Space Center. Paul Weitz and Joseph Kerwin were rookies, about to fly in space for the first time. Conrad was a veteran of both the Gemini and Apollo programs; *Skylab 2* was his fourth space flight. The man in the background assisting the astronauts wears a mask to avoid exposing the crew to possible infections prior to their flight.

Skylab 2 sat on Pad 39B at Kennedy Space Center prior to its launch on May 25, 1973. Charles Conrad, Joseph Kerwin, and Paul Weitz had their work cut out for them; they needed to repair *Skylab*. On blastoff, Conrad told the capsule communicator, "We can fix anything." Seven hours after launch, the disabled *Skylab* was visible. Caught by a metal strap, the solar panel couldn't open. The heat inside the workshop would be unbearable.

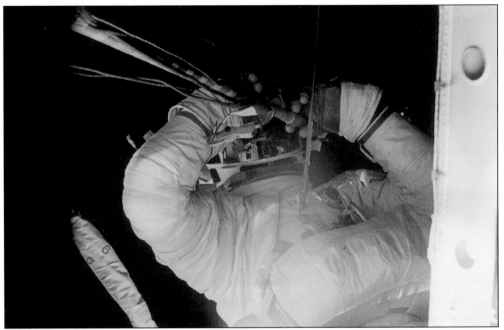

The power problems of *Skylab* worsened on a daily basis. Unless the disabled solar panel could be fixed, the success of the mission would be greatly diminished. A space walk would be necessary to make the needed repairs. On June 7, 1973, Charles Conrad and Joseph Kerwin spent over four hours outside their craft. Joseph Kerwin is pictured attempting to clear the solar panel. The repair was a success; the solar panel fully deployed.

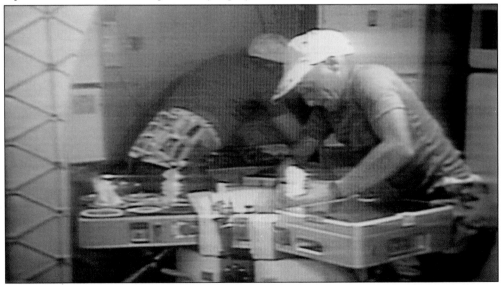

Photographed by a television camera on board, mission commander Charles Conrad and scientist-pilot Joseph Kerwin are pictured in the wardroom of the crew quarters of *Skylab 1*. Compared to earlier missions, the living quarters were magnificent. Provisioned with food, air, water, clothing, and a ton of food, *Skylab* offered sleeping bags, showers, and facilities for recreation. After a spirited game of space darts at 270 miles above the Earth, Kerwin complained, "The darts didn't work worth a darn."

With the success of *Skylab 2*, NASA was ready for another Skylab launch. The *Skylab 3* crew prepared for the mission in the orbital workshop trainer in the mission simulator and training facility at Johnson Space Center in June 1973. Shown here, from left to right, are Owen K. Garriott, Alan L. Bean, and Jack R. Lousma.

As with most missions, scientific experimentation was a major objective. *Skylab 3* was no exception. Besides the crew of three, *Skylab* carried other living matter into space; frog eggs, mice, fruit fly pupae, spiders, and fish were all part of the mission's payload. A bag of North Carolina Mummichog minnows was carried on board in order to subject the fish to weightlessness and to record their level of disorientation.

Skylab 3 sits atop a Saturn 1B rocket at Kennedy Space Center. Launched on July 28, 1973, *Skylab 3* experienced early problems with the thrusters, resulting in preparations for the launch of a rescue craft. Along with mechanical difficulties, the crew suffered several days of nearly debilitating space sickness. Alan Bean, an *Apollo 12* moon walker, told mission control, "We're just not as spry up here right now as we'd like to be."

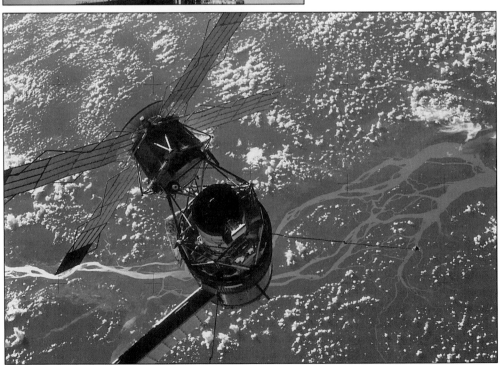

Several hours after launch on July 28, 1973, the three-man crew of *Skylab 3* prepared to dock with *Skylab*. As the 85-ton *Skylab* orbited the Earth at 17,500 miles per hour, *Skylab 3*'s crew photographed the space station against a backdrop of Brazil and the meandering Amazon River. A handheld, 70-millimeter Hasselblad camera, a 100-millimeter lens, and medium-speed Ektachrome film were used to capture this fantastic image.

Skylab 3 scientist-astronaut Owen K. Garriott experienced the thrill of working outside *Skylab* on August 6, 1973. In a record-breaking extravehicular activity (EVA) lasting 6 hours and 31 minutes, he and Jack Lousma worked to deploy the twin solar shield necessary to provide shade for the *Skylab* orbital workshop. In all, *Skylab 3*'s crew performed three EVAs totaling 13 hours and 43 minutes.

Following a journey of 24.5 million miles, the crew of *Skylab 3* splashed down in the Pacific Ocean approximately 230 miles from San Diego, California, on September 25, 1973. After spending 59 days orbiting the Earth 858 times, the crew sat patiently inside the *Skylab 3* command module, waiting to be hoisted aboard the USS *New Orleans*.

The *Skylab 4* prime crew members were: Dr. Edward G. Gibson, scientist-astronaut; Gerald P. Carr, commander; and William R. Pogue, pilot. In a situation unusual for a NASA mission, none of the crew had previous experience in space. *Skylab 4* was planned as the most ambitious American mission to date. NASA planned to keep the *Skylab 4* crew in space for 84 days.

The third and final launch of a manned Skylab mission took place at 9:01 a.m. on November 16, 1973. Hours later, as the crew attempted to dock with the orbiting *Skylab 1*, they experienced difficulty. Finally, on their third attempt, they mated the two craft. Originally scheduled for launch six days earlier, liftoff had to be delayed when an inspection discovered hairline cracks in the tail fins of the aging Saturn 1B rocket.

Highlights of the final Skylab mission were the four extravehicular activities, which totaled 22 hours and 13 minutes. Gerald Carr took this picture of Edward Gibson during the final extravehicular activity on February 3, 1974. Days earlier, the astronauts had searched for the Comet Kohoutek. Spotting the comet, Gibson shouted, "Hey, I see the comet. There's the tail. Holy cow!"

Skylab 4 mission commander Gerald Carr demonstrated the effects of zero gravity as he floated through the forward dome of the *Skylab* space station's orbital workshop. The crew had little free time; each day was spent working on experiments, solar observations, medical activities, housekeeping, and physical training.

Skylab 4 concluded its mission on February 8, 1974, having logged 34.5 million miles and 1,214 Earth orbits. After spending a record 84 days in space, William Pogue, Gerald Carr, and Edward Gibson were able to provide firsthand observations about the effects of long-term space travel. Although they quite successfully performed all of the tasks planned for the mission, the crew of *Skylab 4* suffered from airsickness, poor morale, and bouts of lethargy. They were highly critical of the food, sleeping accommodations, workload, and even the bathroom facilities. Despite the difficulties encountered, however, Pogue and Carr appear to be having a good time here, floating weightlessly as they tend to waste disposal. While Carr holds a bag of trash, Pogue jumps on it, forcing it into *Skylab*'s waste disposal tank. During the three manned Skylab missions, the orbiting workstation had been occupied for a total of 171 days and 13 hours and had been the site of nearly 300 experiments. The unmanned *Skylab* remained in orbit until it fell to Earth on July 11, 1979, scattering debris across the Indian Ocean and a sparsely settled area of western Australia.

Six

A SPIRIT OF
COOPERATION

The Apollo-Soyuz Test Project, first conceived in 1969, took a long time to come to fruition. First, it was necessary to sell the joint Soviet-American space project, the brainchild of NASA director Thomas Paine, to President Richard M. Nixon. Although Nixon had cooled on the idea of space exploration, he warmly embraced Apollo-Soyuz as a means to strengthen détente between the two superpowers. President Richard Nixon and Premier Leonid Brezhnev signed the pact between the Soviet Union and the United States authorizing the project on May 24, 1972. Apollo-Soyuz called for "the docking of a Soviet *Soyuz*-type spacecraft and a United States *Apollo*-type spacecraft." As the first international manned space flight, the project was intended to test compatibility of the different rendezvous and docking systems of the American and Soviet spacecraft. If a rendezvous and docking were successful, it was hoped that there would be additional joint space flights. It was not an easy project; the Soviets were very distrustful of the Americans. Nonetheless, the crew members pose together for this photograph two months before the mission. They are, from left to right, as follows: (seated) Donald K. "Deke" Slayton, American docking module pilot; Vance D. Brand, American command module pilot; and Valerli N. Kubasov, Soviet flight engineer; (standing) Thomas P. Stafford, commander of the American crew; and Aleksei A. Leonov, Soviet command pilot.

A mockup of the Soviet spacecraft at the Cosmonaut Training Center was displayed in Star City, near Moscow. "Everything is ready at the Cosmodrome for the launch of the Soviet spacecraft *Soyuz*," the Soviet Union announced on July 15, 1975, as *Soyuz 19* was launched without a hitch. Meanwhile the American astronauts were all awake in anticipation of their own launch. At 9:10 a.m. they were told, "Your friends are upstairs. Right on schedule." That meant their mission was a go. At 3:50 p.m., *Apollo 18* lifted off.

In orbit, *Apollo 18* astronaut Vance Brand tuned in the correct radio frequency for *Soyuz*. In nearly flawless Russian, the American told the Soviet crew that *Apollo 18* was in orbit, "Miy na khoditsya na orbite!" It would take two days to catch up with *Soyuz*. Pictured is the *Soyuz* spacecraft as photographed by the crew of *Apollo 18*.

Apollo 18 could also be seen from Soyuz. Traveling at 17,400 miles per hour, the two Earth-orbiting vehicles were ready to mate. The controllers gave the word, "Moscow is go for docking. Houston is go for docking. It's up to you guys. Have fun." Thomas Stafford headed for Soyuz. On July 17, the connection was made. Only 52 hours after the Soviet launch, the two spacecraft were joined.

A 35-millimeter camera in the docking module captured this image of American astronaut Vance Brand as he moved through the hatchway from the command module to the docking module of Apollo 18. The historic moment was nearly at hand. Never before had spacecraft from two ideologically opposed nations joined together in space.

A NASA artist depicted the joining of the American and Soviet spaceships during the joint Apollo-Soyuz Test Project. Conversations between the American astronauts and Soviet cosmonauts documented the event as *Apollo* and *Soyuz* docked high above the Earth. "We have capture," said Thomas Stafford on docking. "Well done, Tom. It was a good show. *Soyuz* and *Apollo* are shaking hands," said the Soviets. Soon, the Americans were literally shaking hands with the Soviet cosmonauts. The hatches of both *Apollo* and *Soyuz* were opened. Stafford shook

hands with Aleksei Leonov, after which he and Donald Slayton floated into *Soyuz*'s cabin. A celebration soon followed. Two days later, after being linked together for 47 hours and 96 orbits, it was time to separate. As he powered the *Soyuz* command module away from *Apollo*, Leonov said, "Mission accomplished." *Apollo 18* splashed down on July 24, 1975. The Soviet cosmonauts had returned to Earth three days earlier.

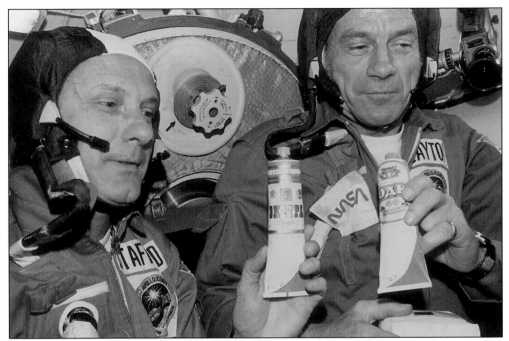

The primary objective of the Apollo-Soyuz Test Project had been to ascertain whether the two separately designed rendezvous and docking systems were compatible. That objective had been achieved. In this picture, Thomas Stafford and Donald Slayton are shown toasting their Soviet hosts with containers of Soviet space food. Although the labels indicated that the containers held vodka, they were really filled with borscht (beet soup).

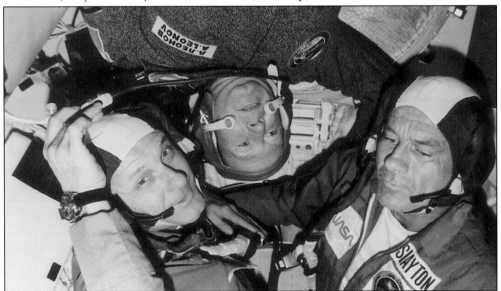

A 35-millimeter camera photographed astronaut Donald Slayton, cosmonaut Aleksei Leonov, and astronaut Thomas Stafford together in the Soviet orbiter module. Despite feelings of distrust between the space travelers from competing nations, a sense of camaraderie was displayed following the successful docking. Establishing détente between the two superpowers was more difficult, however. The Cold War continued.

Seven

SPACE SHUTTLE
1981–1986

With the exception of *Enterprise*, America's space shuttles, *Columbia*, *Challenger*, *Discovery*, *Atlantis*, and *Endeavour*, were all named after sea vessels with significant historical achievements. *Columbia* was the first to fly in an Earth orbit in 1981; it had taken a long time to reach that day, however. The contract for the first space shuttle, *Enterprise*, had been awarded on July 26, 1972. Five years later, *Enterprise* rolled out of the Palmdale, California, Rockwell Air Force Plant 42. By November 1977, *Enterprise*, built strictly as a test vehicle, had largely fulfilled its mission; the shuttle proved that it had the capability of flying in the atmosphere and landing like an airplane. Although contracts for *Columbia* were awarded at the same time as those for *Enterprise*, *Columbia* would not be ready for its first mission for almost nine years. The shuttles *Challenger*, *Discovery*, *Atlantis*, and *Endeavour* followed. As with the Mercury, Gemini, and Apollo projects, there were failures, setbacks, disappointments, tremendous successes, and unthinkable tragedies.

The space shuttle orbiter *Enterprise* rode piggyback atop NASA's Boeing 747 shuttle carrier aircraft during the first free flight of the shuttle approach and landing tests (ALTs) on August 12, 1977. Astronauts Fred W. Haise Jr. and C. Gordon Fullerton crewed the shuttle; Joe Engle and Dick Truly served as the alternate flight crew. The nine-month ALT program took place at NASA's Dryden Flight Research Facility at Edwards Air Force Base.

The 150,000-pound space shuttle orbiter *Enterprise* soared freely only seconds after separation from the NASA 747 carrier. This was the first of five approach and landing tests. Following the flight, the separated aircraft glided to a landing on a dry lake bed. The test simulated landing conditions following the shuttle's return from Earth orbit.

Astronauts Fred Haise and C. Gordon Fullerton are pictured at the controls of *Enterprise* just prior to the fifth and final approach and landing test on October 26, 1977. This test concluded with a landing on Edwards Air Force Base's main concrete runway instead of the dry lake bed previously used. Following several years of additional testing and service as a shuttle simulator, *Enterprise* was given to the Smithsonian Institution.

Riding atop the NASA Boeing 747, the orbiter *Columbia* arrives at Kennedy Space Center on March 24, 1979. Rollout from Rockwell's California facility had taken place only two weeks earlier. Empty weight of the shuttle was 158,289 pounds at rollout. With the main engines installed, the weight increased to 178,000 pounds. *Columbia* was named after a sloop captained by Robert Gray in his 1792 exploration of British Columbia.

STS-1 crew members Robert L. Crippen and John W. Young posed in their ejection escape suits with a model of the *Columbia* space shuttle. Young, one of America's most experienced astronauts, had already logged two Gemini and two Apollo missions. STS-1 would be pilot Robert Crippen's first space flight.

STS-1 *Columbia* was rolled out to Kennedy Space Center's Pad 39A on December 29, 1980. The product of a $9 billion investment, *Columbia* sat on the pad for 105 days in anticipation of its launch date. The first operational space shuttle's launch objective was a two-day check flight culminating in a safe return from orbit and landing at Edwards Air Force Base.

Originally scheduled for April 10, 1981, *Columbia*'s maiden launch was postponed because of computer problems. Two days later, with a tremendous blast of red fire and giant white clouds, *Columbia* was launched at 7:00 a.m. More than a million people witnessed the historic event. Within minutes, *Columbia* reached an altitude of 29 miles and released its pair of reusable, solid-fuel rocket boosters.

STS-1 space shuttle orbiter *Columbia* landed at California's Edwards Air Force Base on April 14. During the flight of 2 days, 6 hours, and 20 minutes, John Young and Robert Crippen orbited the Earth 37 times and flew 1,074,567 miles. At launch, *Columbia* lost 16 heat shield tiles and another 148 were damaged. The first space flight of an American shuttle was an unqualified success nevertheless.

More than 250,000 spectators at the landing site in the Mojave Desert watched as *Columbia* returned to Earth. On touchdown, Houston capsule communicator said, "Welcome home, *Columbia*. Beautiful, beautiful." Following the landing at Edwards Air Force Base, astronaut John Young exited the *Columbia* shuttle and said, "A really fantastic mission from start to finish. The human race is not too far from the stars."

STS-2 astronauts Joe Engle and Richard Truly exited *Columbia* on November 14, 1981. Originally scheduled for a five-day mission, *Columbia* returned early when a fuel cell failure necessitated a premature end to the mission. The shortened flight had not been a total washout; 90 percent of the mission's objectives had been achieved. It was the first time a manned space vehicle had been flown again using a second crew.

Being a shuttle astronaut was not as glamorous as some might believe; hundreds of hours were spent preparing for each mission. Although not nearly as exciting as the actual space flights, extensive training exercises on Earth were a necessary part of the job. Without them, astronauts would have been unprepared for life in space aboard the shuttle. Wearing his extravehicular mobility unit complete with helmet and gloves, C. Gordon Fullerton, STS-3 *Columbia* shuttle astronaut, and another NASA employee appear to be engaged in some sort of karate maneuver or dance step. That was not the case, however. In reality, Fullerton had just completed a suit donning/doffing exercise while experiencing free-fall aboard a KC-135 aircraft. Nicknamed the "vomit comet," for obvious reasons, the plane was specially outfitted to simulate zero gravity and was used to introduce astronauts to conditions they would experience in space.

STS-4 astronaut Henry "Hank" Hartsfield took advantage of the sleeping accommodations aboard shuttle *Columbia*. Hartsfield and his partner Thomas Mattingly were given the mission of putting *Columbia* through its final test. Launch of STS-4 took place on June 27, 1982, with little fanfare or television coverage. The STS-4 mission had the distinction of carrying the first Department of Defense cargo: an infrared and ultraviolet scanner for military spy satellites.

STS-4 *Columbia* returned to Earth on July 4 with the expected patriotic holiday celebration. Following their landing at Edwards Air Force Base, which was viewed by an estimated 500,000 spectators, astronauts Thomas Mattingly and Henry Hartsfield received "welcome home" congratulations from President Ronald W. Reagan and First Lady Nancy Reagan.

The sixth shuttle mission saw the first launch of the shuttle *Challenger*. STS-6, which included a four-man crew made up of Paul J. Weitz, Karol J. Bobko, Donald H. Peterson, and Dr. F. Story Musgrave, was originally scheduled for January 20, 1983. Postponed because of a hydrogen leak, *Challenger*'s launch was delayed until April 4. The five-day mission included a four-hour space walk by Peterson and Musgrave, the first such activity in nine years.

STS-7 orbiter *Challenger* was pictured in Earth orbit during a six-day mission in June 1983. The second flight of *Challenger*, launched on June 18, carried a crew of five: Robert Crippen, John Fabian, Frederick Hauck, Norman Thagard, and Sally Ride. STS-7 traveled a distance of 2.5 million miles and logged 97 Earth orbits. After deploying several satellites and conducting numerous experiments, *Challenger* landed at Edwards Air Force Base on June 24.

Aboard the *Challenger* during the STS-7 mission in June 1983, Sally K. Ride stands on the mid-deck near the continuous flow electrophoresis system experiment. The STS-7 mission will forever be known as the journey during which America's first woman astronaut finally experienced the thrill of space flight. The publicity surrounding Ride's selection as the first American woman to enter space was tremendous. She responded by saying, "I didn't come into this program to be the first woman in space. I came in to get a chance to fly in space." Selected as an astronaut candidate by NASA in January 1978, she brought excellent educational credentials to the program; she held multiple degrees from Stanford University, including a Ph.D. in physics. Serving as a mission specialist on STS-7 and STS-41G, Ride logged almost 350 hours in space. Assigned as a mission specialist on STS-61M, she ended flight training in order to serve as a member of the presidential commission on the space shuttle *Challenger* accident. Ride left NASA in 1987.

Shuttle STS-8 *Challenger* lifted off from Kennedy Space Center's Pad 39A at 2:32 a.m. on August 30, 1983. Following a 17-minute hold for weather, Richard H. Truly, Daniel C. Brandenstein, Dale A. Gardner, William F. Thornton, and Guion S. Bluford went into history as part of the eighth shuttle flight and the first night launching. More significantly, mission specialist Guion Bluford, pictured, became the first African American to venture into space.

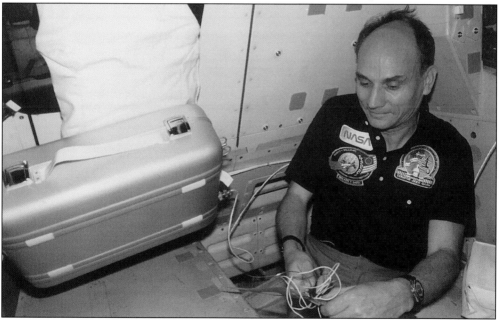

Between August 30, 1983, and April 12, 1985, there were eight shuttle missions. Space flight was considered routine. On STS-51D, launched on April 12, 1985, U.S. Sen. E.J. "Jake" Garn of Utah served as a member of the shuttle crew. Pictured plugging in a food warmer on the mid-deck area of *Discovery*, the 52-year-old Garn was chairman of a Senate subcommittee charged with oversight of NASA's budget.

Four shuttle flights after Sen. Jake Garn's flight, *Atlantis* made its maiden trip. Launched on October 3, 1985, STS-51J was America's 21st shuttle mission. The crew of five, commanded by three-time shuttle flier Karol J. Bobko, logged 1,725,000 miles during a four-day flight. After 64 Earth orbits, the first mission of shuttle *Atlantis* successfully concluded at Edwards Air Force Base on October 7.

U.S. Rep. William Nelson of Florida, payload specialist on STS-61C *Columbia*, prepares to eat a Florida Indian River grapefruit as the shuttle orbits the Earth. Launched on January 12, 1986, the mission was originally scheduled for landing at Kennedy Space Center on January 17. Because of bad weather in Florida, the landing took place at Edwards Air Force Base on January 18.

Eight

CHALLENGER AND BEYOND

The prime crew of STS-51L consisted of, from left to right, the following: (front row) Michael J. Smith, Francis R. Scobee, and Ronald E. McNair; (back row) Ellison S. Onizuka, Sharon Christa McAuliffe, Gregory Jarvis, and Judith A. Resnik. Prior to the launch of the STS-51L mission, *Challenger* had a distinguished history. The shuttle *Challenger* had flown millions of miles in nine successful missions, logged 69 days in space, and orbited the Earth 987 times. Its namesake had been the HMS *Challenger*, a British ship that had sailed the Atlantic and Pacific Oceans in the 1870s. This mission was to include deployment of the TDRS-B satellite and a full slate of experiments, with landing at the Kennedy Space Center scheduled for 144 hours and 34 minutes after launch. When the *Challenger* was launched on January 28, 1986, millions of Americans stared at their televisions in disbelief. The unthinkable had happened.

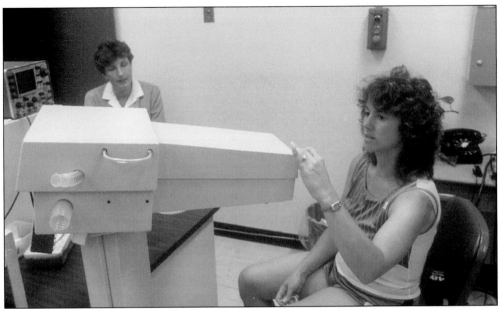

Sharon Christa McAuliffe, a member of the STS-51L shuttle *Challenger* crew, prepares to test her lung capacity in a medical evaluation at the Johnson Space Center. Much publicized as the first private American citizen selected to go into space, McAuliffe was a part of NASA's Teacher in Space project. The third day of STS-51L was scheduled to be McAuliffe's day to shine; she was scheduled to make live teaching telecasts from *Challenger*.

Although a citizen observer and payload specialist, Christa McAuliffe still had to undergo the same types of training exercises as the other astronauts did. Keith Meyers, of the *New York Times,* recorded this image as McAuliffe experienced the effects of microgravity aboard NASA's special KC-135 in a training flight just weeks prior to launch.

Florida's weather during the last week of January 1986 was uncharacteristically cold. As dawn broke on the morning of the STS-51L *Challenger* liftoff, a buildup of ice was clearly evident on Kennedy Space Center's Pad 39B. Although the weather was cold, it was a clear sunny Florida day. Thousands of people at Cape Canaveral thought it would be a perfect day to watch a space launch. The mission had originally been scheduled for a January 22 launch. Delays experienced by mission STS-61C, as well as inclement weather and mechanical problems, had already caused several postponements. Finally, everything came together; the weather was clear, and the shuttle was ready. STS-51L shuttle *Challenger* lifted off Pad 39B at 11:38 a.m. on January 28, 1986.

Almost immediately after launch, things went wrong with *Challenger*. Not even a second into the flight, a puff of gray smoke spurted from the aft field joint on the right solid rocket booster. Additional puffs of yet darker smoke puffed upwards from the joint. The joint had not completely sealed; hot propellant gases were burning grease, joint insulation, and rubber O-rings. Less than a minute later, an expanding ball of gas from the external tank was visible. As thousands of people at the Kennedy Space Center and on the nearby highways and beaches watched apprehensively, it was obvious even to the average spectator that something was wrong. On Florida's West Coast, nearly 150 miles away, the erratic trail of smoke in the sky to the east provided an indication of the unfolding disaster.

Flight directors in mission control monitored the progress of *Challenger* on January 28. The guidance, navigation, and control systems of *Challenger* countered high altitude wind shears 37 seconds into the launch. An unusual small flame appeared on the aft field joint of the right solid rocket booster at 58.8 seconds. As only milliseconds passed, the flame grew larger. The flame spread as it mixed with hydrogen leaking from the external tank.

At 73 seconds into the flight, *Challenger* had reached an altitude of 46,000 feet. A massive explosion resulted as burning hydrogen streamed from the solid rocket booster. Several large sections of *Challenger* were ejected from a tremendous red fireball. Huge plumes of white smoke signaled the shuttle's destruction. The cause of the explosion was low tech; cold weather contributed to O-ring failure in the right solid rocket booster. Seven lives were lost.

After the *Challenger* tragedy, nearly two years passed before NASA resumed shuttle missions. It was up to the crew of STS-26 *Discovery* to return to space. Pictured here, from left to right, are Frederick H. Hauck, Richard O. Covey, George E. Nelson, David C. Hilmers, and John M. Lounge. *Discovery* was launched on September 29, 1988, to begin a four-day mission that ended in a safe return on October 3.

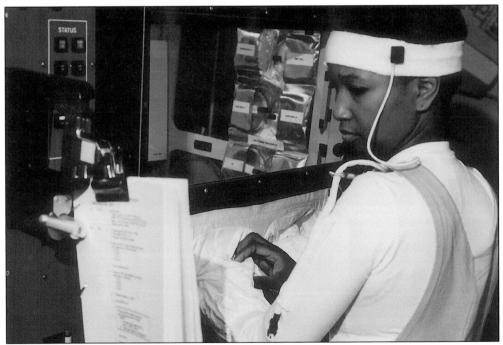

Mae C. Jemison, NASA's first female African-American astronaut, is pictured here aboard STS-47 *Endeavour* in September 1992. Selected as an astronaut candidate by NASA in June 1987, Jemison completed a one-year training and evaluation program that qualified her for assignment as a mission specialist on shuttle flights. A group 12 astronaut and Ph.D. recipient, Jemison logged more than 190 hours in space in her only shuttle mission.

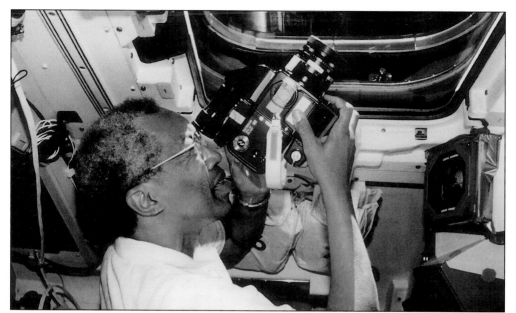

U.S. Air Force Colonel Guion S. Bluford served as a mission specialist on shuttle STS-53 *Discovery* in December 1992. As part of the crew of STS-8 *Challenger* in 1983, Bluford had been the first African American to fly on an American spacecraft. After becoming a NASA astronaut in 1979, Bluford flew on STS-8, STS-61A, STS-39, and STS-53, and logged approximately 700 hours in space.

STS-71 *Atlantis* mission pilot Charles J. Precourt, mission commander Robert L. "Hoot" Gibson, and Kennedy Space Center payload processor Kevin Flautt inspect the Russian-built androgynous peripheral docking system. The mechanism enabled the *Atlantis* shuttle and the Soviet *Mir* to join together as part of the first American shuttle-Soviet space station docking in 1995. When docked, the two craft formed the largest spacecraft to orbit the Earth to date.

At 7:00 a.m. on December 6, 1995, STS-72 shuttle *Endeavour* began the daylong movement on the orbiter transfer vehicle from the vehicle assembly building to Kennedy Space Center's Pad 39B. The ramp to the pad is constructed at a 5 percent incline. Sixty-eight thousand cubic yards of concrete were poured to form the pad's base; its flame trench is 42 feet deep, 450 feet long, and 58 feet wide. Once the shuttle reached the pad, thousands of workers used the

next month to ensure that a detailed checklist of tasks had been accomplished. External fuel tanks needed to be mated with the shuttle; fuel had to be loaded, and navigational systems had to be activated. All manner of equipment such as the fixed service structure, rotating service structure, mobile launcher platform, and orbiter access arm are used to provide necessary ground support functions.

Winston E. Scott, one of Florida's own, was born on August 6, 1950, in Miami. A graduate of Florida State University, he earned a master of science degree in aeronautical engineering from the U.S. Naval Postgraduate School. He logged more than 4,000 flying hours in military and civilian aircraft. Scott served as a mission specialist on STS-72 and STS-87, making three space walks totaling 19 hours and 26 minutes.

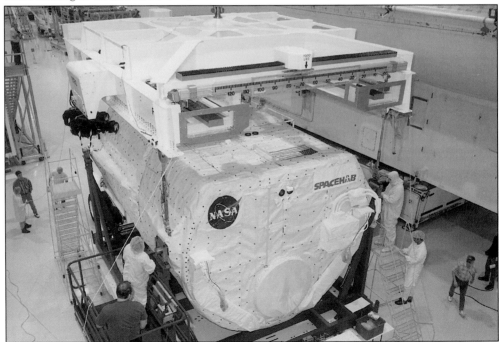

The importance of NASA ground-support teams cannot be overestimated. Thousands of men and women work to ensure a safe and successful flight for each shuttle mission. A team of engineers at Kennedy Space Center inspected the Spacehab-DM double module carried on STS-81 in January 1997. This double module brought thousands of pounds of food, water, and other necessary materials to the Russian *Mir* space station.

John H. Glenn Jr. arrives at Kennedy Space Center three days before the scheduled launch of STS-95 shuttle *Discovery*. As America's oldest astronaut, Glenn, an Ohio senator, had lobbied long and hard for the opportunity to return to space. NASA justified his inclusion as a payload specialist because of his participation in medical experiments on aging. For unknown reasons, however, NASA dropped Glenn from that particular experiment only months before the launch.

John Glenn poses with family members at Kennedy Space Center in October 1998. NASA came under intense scrutiny and criticism for STS-95. Glenn's controversial involvement was described derisively as "a return to the entertainment business" and "very risky." In fact, the risk was minimal. Statisticians computed the risk of a catastrophe as only 1 in 145.

On the morning of October 29, 1998, the members of shuttle mission STS-95 had their flight suits inspected by ground crew Danny Wyatt, Chris Meinert, and Travis Thompson in the environmental chamber. Called the "John Glenn shuttle flight" and the "most expensive congressional junket ever," the hype surrounding STS-95 was incredible. For the most part, the other six crew members were relegated to supporting roles, at least as far as the press was

concerned. Headed for the launch pad, mission commander Curtis L. Brown Jr., far right, led the entire crew of STS-95 from the operations and checkout building. Shortly afterward, each flight crew member entered the shuttle *Discovery* and prepared for a 2:00 p.m. launch. *Discovery's* hatch was closed at 12:30 p.m. and the countdown continued.

Approximately 300,000 tourists and almost 4,000 reporters at Cape Canaveral watched as shuttle *Discovery* soared upward, leaving plumes of white smoke in its wake. The politics and criticism of the past year disappeared as Americans wished for only one thing—the safe return of STS-95 crew Curtis L. Brown Jr., Steven W. Lindsey, Chiaki Mukai, Stephen K. Robinson, Pedro Duque, Scott E. Parazynski, and John H. Glenn Jr.

After almost nine days in Earth orbit, *Discovery*'s crew prepared to return to Earth. Traveling at Mach 7, the shuttle was nearly home by 11:53 a.m. As *Discovery* roared just below the speed of sound over the Gulf of Mexico, still at an altitude of 8 miles, a pair of sonic booms exploded in the atmosphere at midday. Gliding over Kennedy Space Center's 3-mile-long Runway 33, Curtis Brown and Steven Lindsey flawlessly set the 200,000-pound shuttle down at 12:04 p.m.

Following the landing of shuttle *Discovery*, the crew received an enthusiastic welcome from NASA staff members. Payload specialist John Glenn is pictured shaking hands with NASA administrator Daniel S. Goldin. STS-95, the 92nd shuttle mission and the 25th flight of shuttle *Discovery*, was a tremendous success from an operational as well as public relations standpoint.

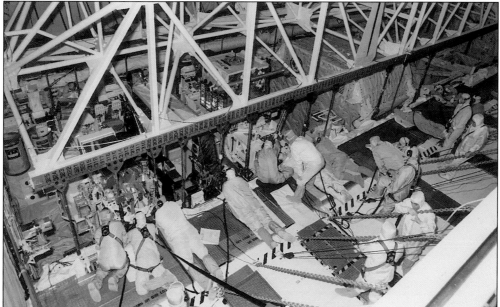

Following the successful nine-day mission of STS-95, NASA technicians prepare to off-load cargo from *Discovery*'s payload bay. During the mission, the crew conducted nearly 100 scientific experiments for NASA, the Japanese Space Agency, the European Space Agency, and several private companies. Equipment carried aboard *Discovery* included the Spartan free-flyer solar-observing deployable spacecraft and the pressurized Spacehab single module.

The prime crew of *Columbia*, STS-93, included, from left to right, the following: (front row) Eileen M. Collins and Michael Tognini; (back row) Steven A. Hawley, Jeffery S. Ashby, and Catherine G. Coleman. Launched on July 23, 1999, the mission objective was the deployment of the 45-foot-long Chandra X-ray observatory. With this flight, U.S. Air Force Colonel Eileen Collins became NASA's first female mission commander.

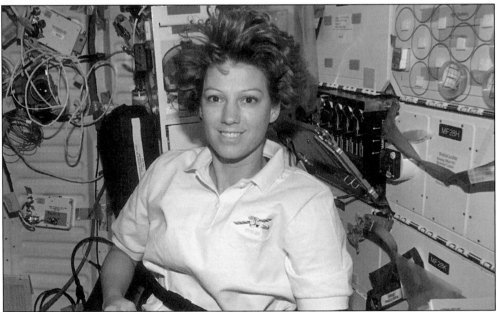

STS-93 mission commander Eileen Collins aboard *Columbia*'s mid-deck loads film into her camera. Selected as an astronaut candidate in January 1990, the highly qualified Collins had over 5,000 flight hours and had served as an instructor pilot, as well as a C-141 aircraft commander. A veteran of two earlier space missions, she believed, "You've got to have the attitude that, 'I am confident to handle anything.'"